Science Talk

Science Talk

Changing Notions of Science in American Popular Culture

DANIEL PATRICK THURS

RUTGERS UNIVERSITY PRESS
New Brunswick, New Jersey, and London

Library of Congress Cataloging-in-Publication Data

Thurs, Daniel Patrick.
  Science talk : changing notions of science in American culture / Daniel Patrick
Thurs.
    p. cm.
  Includes bibliographical references and index.
  ISBN 978-0-8135-4073-3 (harcover : alk. paper) — ISBN 978-0-8135-4074-0
(pbk. : alk. paper)
  1. Science—History.   2. Science—United States.   I. Title.
  Q125.T574 2007
  509—dc22   2006035927

A British Cataloging-in-Publication record for this book is available
from the British Library

Manufactured in the United States of America

To Evelyn Thurs

Contents

Acknowledgments

As I labored in the manic final days of writing this manuscript, I became acutely aware of those whose generosity made my efforts possible. Professor Ronald L. Numbers freely offered his comments and encouragement. His enthusiasm helped me more than once to forget my own doubts and his example as an author increased my respect for the craft of writing. Professor Lynn K. Nyhart challenged me to think more clearly about my work and its implications. She has provided me with a model of the dedicated scholar. Professor Victor L. Hilts was always willing to share his ideas and his time. I would like to think of us as kindred spirits. I must also recognize Professors Craig Werner and James Baughman for their patience, their time, and their insightful feedback.

I am also deeply grateful to the members of the University of Wisconsin–Madison History of Science Department more generally, including my fellow graduate students. Craig McConnell exerted a particularly important influence on my work and my attitude toward being a historian. I am also indebted to Brewer Stouffer, who shared with me the first few tumultuous years of graduate school. My deepest thanks to Chucho Alvarado, Karen Walloch, Hsiu-yun Wang, Rebecca Kinraide, Steve Wald, Jonathan Seitz, and Libbie Freed. I will miss the food almost as much as the friendship. I will not soon forget the companionship of Neil Andrews and Brent Ruswick. Eileen Ward, who repeatedly went out of her way to help me, deserves my profound gratitude as well. More recently, I have also received generous support from the Center for Nanoscale Systems and the Science and Technology Studies Department at Cornell University. I would in particular like to thank Steven Hilgartner.

Without my friends and family, I could never have written this book and preserved my sanity at the same time. Harlan and Jennifer Bjornstad provided me with more than a few welcome evenings of food and conversation. I have for most of my life treasured Harlan's enduring friendship. For all of my life, I have relied on the love and encouragement of my mother, Evelyn Thurs. The value she placed on education has seen me through my graduate career. My sister Robin Kennedy, who got here first, has been a beacon of strength and persistence. My sister Margaret Thurs has been more of a friend than I could have asked. Last, but far from least, I can never fully express my thanks to Carrie Rothburd, one to whom there is no equal. Her support buoyed me up in the stormiest of seas. Her companionship is a limitless source of inspiration. You are all in every page and every word.

Science Talk

Introduction

Talking about Science

MODERN SCIENCE SEEMS TO suffer from a paradox. Numerous observers have noted "the awesome authority that science possesses" in the western world.[1] Sociologists Barry Barnes and David Edge have claimed that "in modern societies, science is next to being *the* source of cognitive authority."[2] Simply labeling a piece of information scientific has often commanded attention and respect, if not assent. Science has, by most accounts, become an especially powerful incantation in American popular culture, even to the point of inspiring supposedly "childlike faith."[3] As early as the 1920s, journalist Frederick Lewis Allen claimed that in the minds of the "man in the street and the woman in the kitchen" the "prestige of science was colossal."[4] In a variety of surveys, Americans have consistently expressed their positive views of science as an engine of progress and a force for good in the world.[5] Popular movements, such as those that arose around eugenics and public health reform during the early 1900s, have occasionally clothed themselves in scientific garb. And American consumers have routinely demonstrated their enthusiasm for science-enabled technologies. By the closing decade of the twentieth century, anthropologist Christopher Toumey reflected that the symbolic power of science as a means to "answer any of life's questions" was so great that American citizens respected it "as a kind of religion."[6]

It hardly seems any wonder then that many modern science watchers have concluded that modern science has "become too important to be ignored, even by those who do not understand it or who reject it." And yet, Americans seem to have found ways to ignore it easily enough.[7] Philosopher

Paul Kurtz asserted during the mid-1990s that science had in fact "lost considerable" prestige since a golden age of popular appreciation during "the nineteenth and the first half of the twentieth centuries."[8] Recent media coverage of second-hand smoke, for instance, gave science "only a secondary and often relatively minor supporting role in the story" despite the availability of scientific research on the subject.[9] Instead, journalists were turning elsewhere for compelling storylines. Many consumers of information have turned elsewhere as well. The 2002 National Science Foundation report on science and engineering indicators concluded that only "2 percent of the most closely followed news stories of the past 15 years were about scientific breakthroughs, research, and exploration."[10] The same polls that revealed Americans' respect for science have found that the average citizen's grasp of scientific knowledge is often fairly feeble.[11] Ultimately, zoologist and popularizer Richard Dawkins found it "bafflingly paradoxical that the United States is by far the world's leading scientific nation while *simultaneously* housing the most scientifically illiterate populace outside the Third World."[12]

Dawkins has had company in his bewilderment. A wide array of scientists, educators, and popularizers have wondered at the seeming disconnect between the critical role of scientific knowledge in the modern, industrialized world and indications that a large segment of the American public was "merely uninterested in, or perhaps bored by, science."[13] Kurtz perceived this situation as similarly "paradoxical."[14] Explanations have included the influence of extra- or antiscientific factors such as religion, political extremism, or superstition; the barrier of attention spans ruthlessly shortened by exposure to entertainment-driven mass media; the failure of those responsible for popularizing scientific knowledge to communicate the essence of "real" science; and the spread of fear or even resentment of scientific power. Such offerings rightly suggest that the average nonscientist's relationship with science is a complex one that has been shaped by a number of factors, from the commercialization of the press to the specialized and often highly abstract nature of present-day scientific information. But the subtle place of science in American culture has also resulted from a fairly simple fact; though science is often taken to indicate more or less directly a set of specific practices or a body of knowledge, it is first and foremost a word. The ways in which we relate to science have a great deal to do with how we have learned to talk about it. In particular, descriptions of science that have infused it with power and authority have, when given a slight twist, also depicted it as a subject that could be safely ignored. This duality has become one foundation for the "paradoxical" place of science in American culture.

Present-day Americans regularly depict science as something unlike other kinds of knowledge or practices by marking it off as one-half of the binary division between "science and religion," identifying it by reference to a special kind of "scientific method," or locating it in the actions and statements of a unique group of highly trained "scientists." These distinctions have opened the door to attestations of the value of science in contrast to other kinds of information. Science is special because it is free of the human foibles, subjectivity, and irrationalities evident elsewhere. It can accomplish incredible feats, reach into regions unimagined by the ordinary person, and reveal facts and ideas that are well beyond the typical mind. But a science more easily set apart has also been a science more easily set aside; greater distinctness has created novel possibilities for subversion and containment as well as celebration. Sociologist Mike Michael has examined what he called "discourses of ignorance"—strategies that interviewees outside of the scientific community have used to explain and justify their own avoidance of science. These included "a bracketing of scientific knowledge as 'other'" in the sense that the less a person understood a given subject, "the more science-like it is."[15] People also relied on a supposed division of labor between those whose job it was to understand science and the rest of the population, opening up the space for interviewees to claim, in Michael's words, that "I am *not required* to know this stuff" or that "*I don't need* to know this stuff."[16] Finally, ignorance was sometimes presented as an explicit choice, either because of the perceived irrelevancy of detailed scientific information or because of an "effort to maintain social independence from science and, possibly, to challenge the authority of interests" claiming to speak in its name.[17]

While American's apparent ability to sidestep the scientific has often been presented as a failure to understand the nature of "real" science, avoidable if only there had been better education, popularization, or public relations, the ways people talk suggest a much more complex and ambiguous relationship with the scientific as a category. The very same language that public commentators used to laud science as an independent and prestigious brand of knowledge could simultaneously make the scientific remote, raise challenges to depicting science or its methods as broadly relevant to ordinary life, and justify rejection, resistance, or even ignorance. Of course, modern habits of depicting science did not come from nowhere. They have a history and, in the following chapters, I will be following that history through a number of high-profile public debates over the scientific status of certain ideas, including phrenology, evolution, relativity, and UFOs. Each controversy pushed people to define what they meant by science or to dispute the definitions of others. They also provided crucibles where new conceptions of the scientific sometimes formed. I will conclude by tracing the themes that emerge from this story as they have appeared amidst present-day discussion of intelligent design. Finally, I will take the opportunity of the intelligent design debate to reflect on some of the possible responses to the "paradox" posed by modern talk about science.

The Power of Words

Science as a word has a long story. Its meaning has changed dramatically over the last two centuries. The term entered English from French during the Middle Ages to denote reliable and systematized knowledge. It had no special connection to nature or to the physical sphere and was often more or less synonymous with philosophy. This traditional meaning persisted into the early 1800s. The 1806 *Webster's Dictionary* simply listed science as "knowledge, deep learning, skill, art." The 1850 edition indicated that it meant general or philosophical knowledge. Very little popular rhetoric distinguished science, while the vast bulk of commonly offered depictions emphasized what it shared with complementary truth-seeking activities. Between the late 1800s and early 1900s, however, new ways of talking joined traditional ones. Instead of demonstrating the continuous place of science in general culture and the search for truth, these depictions created distinctions between science and other realms. By the end of the twentieth century, the equation between science and general knowledge was all but lost. "Dropping science" means to communicate some important information in the vocabulary of modern rap lyrics.[18] Nearly everywhere else they described its boundaries, Americans portrayed science as distinct from all other human activities.

The rhetorical foundations of the authority of scientific knowledge changed over the same period of time. During the eighteenth century, many scientific pronouncements gained their force from the appearance of accessibility and openness. Likewise, the frequently amorphous science of the early 1800s was more or less transparent, important not in itself but because it was part of a harmonious whole and a vehicle for general truth about the world, natural and otherwise. Increasingly opaque and impenetrable boundaries around science interrupted its continuity with other realms and, at times, made science the focus of attention rather than the world of truth behind it. The cultural prestige of science flowed more and more from its comparisons and contrasts with other sorts of information about the world. The key rhetorical issue in many twentieth-century public scientific controversies has been whether new ideas were scientific or unscientific, not directly whether they were true or untrue. But the growing distinctness also meant that the general intellectual value of science, previously derived from its connections with other areas of thought, could be severely limited. At times, trying to assert its transcendent importance has violated the very boundaries between science and philosophy, religion, or other examples of not-science that made the scientific uniquely worth paying attention to in the first place. Most of the time, only technological application remained to connect the walled-in scientific world and the rest of Western culture.

Such an account is the story of a word, but not *just* a word. The American science fiction author Philip K. Dick once wrote that "the basic tool for the manipulation of reality is the manipulation of words."[19] There is certainly a universe out there that is aloof from what we say about it and words refer to aspects of that universe. Those who have talked about science over the last several hundred years have claimed to make reference to real bodies of knowledge or sets of practices, changes in which have been reflected in the evolving meanings of the term. Knowledge frequently associated with science has become more specialized, technical, and often mathematical in response to some difficult problems posed by natural phenomena. That growing complexity alone has had a role in setting science apart from other kinds of information, particularly the much less formalized sorts that most Americans have learned to deal with on a daily basis. Likewise, scientific practice has undergone what historians routinely call professionalization.[20] Between the late 1800s and early 1900s, a variety of institutional structures emerged that set researchers in scientific fields apart as a professional class, protected their autonomy, determined correct procedure, and moderated disputes by sanctioning some kinds of knowledge as real science. The organization and formalization of scientific practice naturally contributed to depictions of significant differences between the methods of science and those used in other areas of human endeavor and provided a basis for distinguishing between those who pursued science and other kinds of individuals and groups.

However, the ways in which people talked about science were not only reactions. Like many of the important categories that we use to divide up what we know, perceive, and do into manageable pieces, science has been created as much as observed. Nor have the purely rhetorical dimensions of categories such as science been merely add-ons. Simply by being spoken, they have called attention to certain aspects of the world while allowing others to be safely ignored, made particular courses of action or trains of thought more convenient and obvious than alternatives, and established the basis for the generation of cultural authority.[21] In fact, ideas about the nature of science were often instrumental in creating the things to which it has referred. The trend toward more specialized scientific knowledge was in part made possible by shifts in the perceived purpose of science away from enlightening the masses and toward understanding and manipulating nature for its own sake. Over the same period of time, aspiring professionals were among the most enthusiastic purveyors or highly bounded visions of science, even before the social organization of scientific work was particularly secure. Professionalization was as much a rhetorical exercise as a social one. Forming a closed group of practitioners who could speak for science was hampered so long as that science seemed diffuse and without strict borders. Distinct images of science, whatever their details, provided a flag around which professionalizers could rally and the means to exclude those who did not belong.

The history of science as a category has also been much more than a transition from one dictionary-style definition to another. At any given moment, science has been the bearer of varied, sometimes even contradictory meanings.[22] There is nothing here unique to science. As Oliver Wendell Holmes noted, "A word is not a crystal, transparent and unchanged; it is the skin of a living thought and may vary greatly in color and content according to the circumstances and the time in which it is used."[23] That is particularly true of the most widely shared keywords in any culture, the terms that provide guideposts to which people can refer as they navigate the events of their lives.[24] In the modern United States, prominent keywords include "family," "race," "freedom," and "America," as well as science. Though they often give the appearance of being self-evident, these words have all had complex histories that have resulted in deep reservoirs of alternate meanings. Of course, the openness of keywords to multiple interpretations has meant that they are frequent subjects of misapprehension and debate. But in the absence of questioning, that same flexibility can bring disparate groups and individuals together by at least allowing them to believe, like a well-placed wink, that they are talking about the same thing. The many potential meanings associated with keywords have ultimately made them enormously powerful vehicles for cultural authority, adaptable to many tasks, and suitable for achieving a variety of aims. Such power is what makes them worth fighting over in the first place

The close relationship between cultural authority and flexible meaning has been revealed with particular clarity by the notion of boundary-work.[25] Scientific practitioners have engaged in boundary-work when they defined science in ways that enhanced its value as a special category, ensured their own access to that power, and denied it to their competitors. In a classic example, the nineteenth-century man of science John Tyndall defined the scientific for audiences of mechanics and other practical men in ways that stressed its transcendent, poetic, and cultural value. His statements to theologians tended to emphasize the material and concrete aspects of scientific work and knowledge. In both cases, he could claim special expertise his audiences did not have.[26] Many analyses of boundary-work stop at this point, but there is even more to the story. The members of Tyndall's audience could also define science. Some may have done so in ways that augmented his claims. Others could have depicted science in ways opposite to Tyndall's, thereby denying his possession of special knowledge they did not have. Finally, a few listeners might have accepted Tyndall's assertions about science, but used them to confine his expertise to matters of little or no importance. In this way, the flexibility of rhetoric has been a source of power but also of resistance to that power.

Taken individually, instances of boundary-work tell us relatively little, except that people have adjusted the definition of science in order to suit their own purposes. To get a better sense of the place of science as a keyword in American culture requires that we look past the fact of boundary-work to the basic materials with which boundaries have been made. That is, we need to take stock of the "science talk," by which I mean the general reservoir of boundary-making words, images, and ideas that Americans had on hand when they were called on to define the nature of science. The availability of these rhetorical and conceptual resources and the ways in which people chose to use them shaped their sense of meaning in the same way that the folkloric motifs have influenced the imaginative universe glimpsed through storytelling or that physical tools have given rise to the kinds of objects and structures that fill day-to-day existence. When it was science, rather than a house, that needed constructing, people turned to the tools of contemporary science talk.

Science Talk

The notion of science talk joins a long list of "talks," including God talk, crime talk, and even DJ talk to indicate the between-song banter of radio disc jockeys.[27] All of these have roots in the scholarly concept of discourse.[28] Discourse is the method through which cultural meaning is constructed or, to say it another way, talked into existence. It is a step up in abstraction from the more familiar concept of discussion. A discussion implies a particular conversation about a specific topic conducted by a known group of persons. It may occur over a long period of time, proceed in writing or in speaking, and its focus may evolve from one set of issues to another. Yet a discussion always possesses some degree of concreteness. A discourse frequently transcends particular people, places, and times, though it may be associated with certain socioeconomic groups, geographical zones, or epochs. It is a kind of meta-discussion or a set of templates for a group of discussions linked by a global topic area, a general context, or a keyword, such as crime or the radio airtime between songs. As a result, active discourses provide models for what people in a given culture can possibly say and how they can say it.

But the concept of discourse also allows considerable variation. It limits but does not determine how people will talk.[29] Individual discourses do not amount to a coherent and consistent whole. As I implied above, the raw materials of a given discourse are better imagined as a toolkit of possibly contradictory words, images, and ideas, and the rules for their use.[30] People can use these tools to construct the things they talk about in particular discussions in the same way that they put tools to work to make a specific object. We do not require that hammers and screwdrivers be in any sense consistent with one another as long as they do what we want them to do and the individual structures they create are sound enough to be useful, at least for as long as they are needed. Likewise, there was no requirement that the rhetorical science-defining tools widely available, once gathered together, would assemble a unified image of science. At the same moment in time, science has been depicted primarily as both a collection of facts and a matter of theories, a source of transcendent truths and new gadgets, the possession of all thinking people and scientists alone. Each of these assertions does certain kinds of discursive work. All are equally part of science talk.

The contents of science talk have tended to cluster into a number of specific genres. Methodology has long been one. Another has involved explicit juxtapositions of science and other cultural categories, whether technology, philosophy, or religion. The history of science has provided an additional means of demonstrating the true nature of scientific knowledge and a stage on which to compare it with other sorts of enterprises. Finally, the ways that people have talked about those who did science became intertwined with talk about science itself. Claims about method could sometimes be abstract and prescriptive, but they also turned into depictions of what scientific practitioners really did (or did not do). Likewise, differences between kinds of knowledge were often expressed as differences between the sorts of people who were seen to represent them. Debate about whether scientists were like or distinct from ordinary human beings turned into a major vehicle for working out the relationship between scientific and common knowledge. Particular historical figures, such as Isaac Newton or Francis Bacon, also appeared as central characters in stories about the development of science.

Centering our analysis on rhetorical tools helps us to place boundary-work around science in its broader social, cultural, and historical contexts. In scholarly practice, the capacity to engage in boundary-work has generally been depicted as the sole possession of professional scientific practitioners intent on asserting the value and autonomy of their field. Because it involves tracking the use of special words and phrases rather than particular groups or motivations, following science talk encourages looking outside the community of professional scientists and beyond attempts to make science important. Trends in science talk also tell us about changes in the ways scientific boundaries were made. Whole genres of rhetorical resources have risen or declined in importance. For most Anglo-Americans, historical anecdotes were a widely shared means of making sense of science during the 1800s but have become somewhat more rare in modern discourse. There were also shifts within species of science-defining tools. Certain words, such as "scientific method," became more or less popular over time. The intensity of attention to comparisons and contrasts between science and philosophy, religion, common sense, and technology also varied in ways that mirrored contemporary concerns. Looking at boundary-work as a kind of strategy without stressing its content can lead to a more episodic and fragmented picture in which the fact of boundary-making is identified in historical debates, but not put together with other examples into a more general story.

Other ways of making sense of science in popular culture have sacrificed the recognition of flexibility in favor of coherence and convenience. Studies of science in the mass media have tended to rely on an assumption that images appearing in newspapers and magazines became "embedded in popular consciousness," sometimes in the form of a larger "worldview" that reflected the beliefs of social and cultural elites.[31] Such a characterization, sometimes justified by appeal to an "American democracy, with a free press and marketplace sensitivity" in which "commercial success and audience acceptance go hand in hand," has allowed readily available mass media sources to serve as easily accessible stand-ins for the much less accessible popular mind.[32] But it also makes readers essentially passive recipients. This passivity has reached its purest form in the notion of "memes," which were first discussed by Richard Dawkins in *The Selfish Gene* (1976). Memes, including widespread words and slogans, supposedly represent the basic building blocks of ideas, just as genes do for biological organisms. In that way they resemble the basic tools of science talk. Yet memes are also imagined to replicate themselves, like viruses. The ordinary person is no more empowered by them than by the cold they catch from a friend; rather, memes eliminate the middlemen of the mass media and replicate in the brains to which they are exposed of their own accord. Under the guise of so-called "memetics," human beings simply become machines for transmitting memes. Though memetics has as yet exerted little or no influence on detailed histories of culture, it has achieved some popularity outside of professional scholarly circles, particularly in discussion of the Internet. Nevertheless, it ultimately confuses the distinction between the relative stability of physical objects, such as sequences of genetic code, and flexibility and messiness of rhetoric.[33]

Attempts to turn away from the ambiguity of words have in fact helped to make the "paradoxical" place of science in modern American culture appear so paradoxical, often by focusing attention either on slavish acceptance or irrational rejection.[34] A "popular mind" that believed everything it heard on the news or was trapped in a prisonlike worldview that it had absorbed from the classroom, the pulpit, or the mass media must either accept science totally and on faith alone or irrationally reject science under the thrall of competing systems of belief, whether religion, pseudo-science, or antiscience. Many histories of popular scientific imagery have taken as their mission explaining how science advanced to its present position of overwhelming authority, thereby making the prestige of science a fact about popular attitudes to be explained rather than a phenomenon to be explored or defended itself. Accounts of boundary-work have in practice become intertwined with accounts of deference to scientists and the promotion of science as an authoritative category. Tyndall's audience has, by and large, been missing. Alternately, many surveys of science in twentieth-century mass media, typically written by those with an interest in or sympathy for science popularization, have striven to explain the lack of cultural authority that "true" science possesses by pointing out, for instance, how journalistic distortion has prevented scientific knowledge from achieving its rightful place in the popular mind. Still, suggestions of a more complex view have found their way to the surface even where depictions of scientific prestige have been most pervasive. A reviewer of historian Marcel LaFollette's *Making Science Our Own*, which traced images of science in twentieth-century American magazines, observed that "on the one hand, the author tries to show that we have become slaves to science and, on the other, she shows how public opinion polls suggest otherwise."[35]

The notion of science talk as I have outlined it above abandons the quest to know what people have really thought about science or to bind individual depictions of science together as manifestations of a semioccult "popular mind." Rather, the only things we have direct access to are utterances about science made at particular moments for specific purposes. Evaluating the spread of various rhetorical tools through such utterances tells us both about the habits of talk that were most common among authors and also the kinds of science talk that many readers encountered, were familiar with, and perhaps added to their personal collections of science-defining words and images. Though it provides some basis for conclusions beyond our immediate sources, however, the general reservoir of science talk accessible through public discussion of science is not another kind of public mind. We cannot say that Americans simply repeated what they read. Recent research on audiences has emphasized that while the mass media do limit people's perception of the world, especially by framing issues and providing the raw materials with which to express impressions, readers have still had considerable freedom in their reaction to texts and in their use of the images that they found there.[36] Likewise, American readers were free to use (or not use) the elements of science talk they gathered in the ways that made the most sense to them. Rather than predicting their exact responses or revealing their hidden thoughts, understanding the general reservoir of rhetorical tools from which they assembled their expressions allows us to construct the most likely ways they might have defined science when they felt called upon to express their themselves, even if those views were never recorded.[37] Most never were. Thus, acknowledging the agency of readers in their use of science talk limits what the modern historian can conclude but offers a much more complete, flexible, and ultimately far-reaching picture of the place of science in American culture.

Science in Context

By putting changes in different kinds of science talk together, we can more easily see larger patterns that reveal the increasing distinctness of science. During the first half of the nineteenth century, the bulk of widespread Anglo-American science talk did little to set science apart. Its methods were the same as those that artists, theologians, and even ordinary people used everyday. Its practitioners were not distinguished as a special class. It shared the same historical trials and tribulations as every other kind of truth. Terms that distinguished science as something special, including now familiar pieces of science talk such as "scientific method," "science and religion," and "scientist," began to circulate widely only during the last quarter of the 1800s. By the early twentieth century, these and other rhetorical tools that were good for making strong boundaries around science dominated older kinds of science talk. They depicted science as distinguished by a wide range of attributes and accomplishments. Discussion of a special scientific method reached new heights. Scientists seemed to be another species of person entirely, sometimes looking down on the ordinary world like demigods, at other times locked away behind the doors of their labs or with their heads in the clouds. In the end, science was not only more easily differentiated from other sorts of knowledge, it was harder and harder to say how it was like them. After mid-century, the otherness of science constructed with commonly available aspects of science talk was unquestionable.

Focusing on science talk helps us grapple with the complex place of science in American culture, but it does not by itself explain why the basic materials of science talk changed or why Americans used them to simultaneously set the scientific apart and aside. Rather we find answers to these questions only when we juxtapose the growing distinctness of science with a trend in American culture at large toward greater boundedness, by which I mean the proliferation of distinct categories for dividing up the world. We do not find increasing distinctness and division everywhere we look back. The contempt that kept the Irish beyond the pale of proper nineteenth-century American society has disappeared. Nevertheless, the multiplication of Irish-Americans, Asian-Americans, African-Americans, and so on suggests that modern-day rhetorical tools for distinguishing between kinds of people have hardly decreased.[38] New divisions arose in the world of information especially. Between the late 1800s and early 1900s, newspapers and magazines began to segment their content into special sections or specialized publications. Over the same period, formulized genres, including science fiction, mystery, and romance, gave order to popular literature, while public libraries began offering these and other sources of information and entertainment to patrons, neatly categorized and arranged.[39] Perhaps no area of American life exploded in quite the same way as consumer goods. By the second quarter of the twentieth century, a massive advertising industry had arisen to divide those goods into brands and infuse those brands with meaning.[40] In many ways, the construction of a well-bounded science was the creation of a powerful new brand of knowledge.[41]

The notion that science has become a brand name as much or more than the bearer of transcendent and ultimate truth has not appealed to everyone. In *How Superstition Won and Science Lost* (1987), historian John Burnham preceded Paul Kurtz in arguing that the high point of science popularization came during the late nineteenth and early twentieth centuries. During these years, as Burnham's tale goes, men of science were filled with zeal for the religion of science and the urge to crusade against error and superstition, and they provided the public with information embedded in a broadly scientific view of the world. Over time, however, scientists left the field of popularization to journalists, whose influence was "altogether corrupting, turning high culture into trivial news items." Meanwhile, scientists lost their status as the heroes and interpreters of modernity to the likes of advertising agents.[42] In such dangerous hands, an increasingly commercialized and fragmented mass media presented scientific ideas and discoveries as isolated facts, thus dismembering the scientific worldview and hobbling the religion of science. The effect was that science took on the form of superstition, which for Burnham meant a disjointed set of laws or beliefs accepted largely on the basis of faith in authority.

Yet Burnham makes little allowance for the possibility that certain ways of talking about science spread, not because of the power of advertisers or audiences' passive acceptance of mass media imagery, but because they were actually useful or even necessary. Most of the changes in the American rhetorical landscape over the last two centuries originated as attempts to exert some control over the increasing scale of life in a quickly industrializing nation. The formation and commercialization of the mass media dramatically expanded the amounts of information available to the ordinary American. Dividing it into pieces helped publishers create audiences and simplify marketing while it gave consumers the capacity to structure their own consumption.[43] Similarly, brand names both enabled people to make choices about what they purchased and provided what advertisers hoped were the means to manipulate those choices. The growing boundedness of science allowed Americans to read or pass over a magazine article or book simply because it was about science without having to know anything more about its content. Rhetorical and social control converged most powerfully in the efforts of certain professional groups to claim realms of newly segmented information for their own. Professional scientific practitioners, including Burnham's crusading scientists, were one product of this convergence. But they were not alone. Between the late 1800s and early 1900s, a wide variety of groups from physicians to teachers to journalists sought to organize themselves and to assert their ownership over particular regions of expertise. Often, aspiring professionals appealed to science as a means of distinguishing between insiders with legitimate knowledge to offer and outsiders who had only flawed or false information at hand. Talk about scientific medicine, engineering science, educational science, and scientific advertising all proliferated in the decades around 1900.[44]

The end result of all this activity was that, by the latter half of the twentieth century, the United States had become a highly bounded information culture filled with experts on topics from international relations to parenting. Yet American public rhetoric has, particularly since the 1820s, also been shot through with celebrations of democracy and the promotion of the people's ability to judge matters of utmost importance. The growth of expertise was a curtailment of that ability and it set limits on the reach of democratic processes by restricting the relevant "people" on certain matters to exclusive groups of highly trained and often socially unrepresentative communities. Scientific truths, as is sometimes repeated, are not decided at the ballot box. Instead, experts occasionally offered a modified form of democratic sentiment in which bearers of expertise offered basic facts to the voting population, which in turn had a right to decide what to do in response to those facts. Still, heeding every possible claim of expertise would have narrowed the ground of popular sovereignty to almost nothing. Nor was it clear what to do when experts disagreed, as they sometimes did. At other times, expertise overflowed its original boundaries as experts sought to annex new areas of experience to the realms for which they spoke. In such a light, the impetus to rhetorically set experts, and the science on which they often based their assertions, aside helped to clear out a space for individual judgment and extend democratic ideals from politics into the world of knowledge.

I do not mean this characterization in any way to justify or ennoble the ability of Americans to ignore science. Rather, I intend only to suggest that popular science talk shares one of the main functions of popular culture more generally, namely, to enable resistance to forces perceived as attempting to dominate from above. From their emergence in early modern Europe amidst growing political and economic cleavages between upper and lower classes to their continued development during the last two hundred years among growing gaps between experts and laypeople, professionals and amateurs, insiders and outsiders, the raw materials of popular culture have remained adept at making fun of the serious and turning the profound on its head, whether through the ridicule of kings or of the pretensions of experts.[45] Communication studies scholar John Fiske has stressed the existence of popular culture in reaction to attempts to exert widespread control and argued in particular that "a text that is to be made into popular culture must . . . contain both the forces of domination and the opportunities to speak against them."[46] The unambiguous cultural authority of science has only really existed in the context of powerful institutions that could not only control the ways in which people spoke, often by encouraging or even demanding the usage of technical jargons, but could also reward certain approved interpretations while censuring others. Indeed, one of the major functions of institutions in the modern day, like the concepts of worldviews and memes, is to control and discipline the messiness of meaning.

Science Talk on the Page

The potential ambiguity and flexibility of keywords such as science suggest that we cannot understand them either by reference to the opinions of a few "great men" who might be assumed to represent the thoughts of a given age or through logical arguments about what people must have imagined based on a subset of their expressions. If we take seriously the active struggle of authors and readers to make sense of science and draw meaning from its discussion, then the line between scholars, whose views are to be engaged rather than attributed to social factors or worldviews, and those they study, whose claims can be explained away at will, should perhaps be blurred. Rather than focus on hidden meanings, available only to the modern scholarly observer, I believe a better approach to science talk comes through treating historical actors not as subjects of analysis but as colleagues in the attempt to discern the meanings of science. That is, wherever possible, we should let people talk for themselves.

In order to trace the popular uses of American science talk, I have relied in particular on widespread rhetoric available in weekly, semimonthly, or monthly magazines.[47] Discussions of science have occupied a prominent place in nontechnical magazines, to an equal if not greater degree than in other kinds of mass communication. Many early-nineteenth-century monthly or weekly periodicals, such as the *Southern Literary Messenger* or *Ladies' Repository*, were highly eclectic in their choice of subject matter and included articles on a wide variety of topics from natural history to mining. Publications linked to specific religious denominations, which flourished in the period, also paid attention to scientific matters. After mid-century, many journals turned more toward the coverage of current events than the discussion of general topics, but they continued to include items about science, often in sections specifically set aside for scientific news. Between the late 1800s and early 1900s, magazines became cheaper and gained higher and higher circulations. Up until the advent of radio, they were the only mass media with any claim to a national scope. Even with the emergence of radio and the early stages of television, magazines remained a widely shared and significant source of information. And science remained a source of articles in their pages, though not an overwhelming one. During the early 1900s, a number of prominent popular magazines averaged one science article in every two or three issues.[48]

In the present day, a significant amount of news about science intended for the general reader also appears in magazines devoted in particular to scientific topics. The development of such scientific periodicals was itself an important step in the growing boundedness of science. During the antebellum years few periodicals limited themselves to scientific information. The *American Journal of Science*, founded in 1818, was among the earliest. After mid-century, however, there was an explosion of agricultural, medical, astronomical, and mechanical publications, among others. Magazines partly or totally focused on science generally also appeared, including *Appleton's Monthly* in 1869 and the *Popular Science Monthly* in 1872. By the early 1900s, the basic pattern of special-interest science publications was well established. Psychologist and publisher James McKeen Cattell's *Scientific Monthly* took the place of the *Popular Science Monthly* in 1915. After World War II, there was a concerted attempt by scientists and science popularizers to revamp or establish a number of general science journals. These publications represented attempts to fix the legitimate boundaries of science for general readers, but they were not alone. Science magazines were only one part of the larger trend toward specialization. The sheer number of different kinds of magazines by the twentieth century offers the historian a greater range of voices than was found in most other media. This diversity allows us to take a fuller measure of widespread American science talk, its patterns, and its complexities.

Magazines also make good places to sample talk about science because they occupied a unique spot on the cultural landscape. In the mid-1830s, Edgar Allen Poe declared that the "whole tendency of the age is Magazineward."[49] Insofar as readership was concerned, he exaggerated. Then and now, magazine circulation has tended to be largest among those with higher socioeconomic status and more education. But magazines have consistently occupied a strategic position between books, which often involved substantial investments of time and sometimes money, and more ephemeral sources of information, such as newspapers and lectures. Especially during the early 1800s, when the boundaries between different kinds of publications were fairly permeable, magazines often reprinted items from newspapers, which frequently borrowed among themselves, or reprinted lectures, which were often heard by large numbers of people. Even in the twentieth century, when a much higher percentage of magazine content was produced specifically for that magazine, articles sometimes referenced stories in newspapers, programs on the radio or television, or films. When I have relied on other sources of science talk in this study, I have put special emphasis on those popular lectures, books, or programs that were cited, reviewed, or otherwise noticed from the vantage point of periodical literature.

In terms of coverage, magazines balanced extended discussion of science with readability. They have been both a rich source of rhetorical tools and one that large numbers of readers were likely to encounter. The sometimes opinionated orientation of many magazines also provided frequent encouragement to assertions about the nature of science. Throughout the nineteenth and twentieth centuries, their contents have not simply reported facts, but have also been intended to evaluate the topics of the day and ultimately to "influence belief and conduct."[50] "Science-in-context" stories, which sought to show the importance of the latest discoveries and their implications in a broader perspective, appeared earlier in magazines than in other kinds of mass media.[51] Their success as determiners of popular ideas is open to substantial debate. But the pages of magazines often became sites of controversy and struggle over matters political, economic, scientific, and more. Opposing voices sometimes sounded in the same periodicals or, as the number of magazines grew, in competing publications. Articles from different magazines frequently took issue with one another. Again, all of this provides a rich soil from which to harvest examples of science talk.

Magazines' tendency to take up partisan positions amidst debate is a particularly important advantage. To discover living, active talk about science, we need to focus on situations that prompted people to use it. Controversies, particularly popular ones, provide the best place to watch Americans putting their science-defining vocabularies to work in ways intended to be seen by others. The most intense controversies could also spark the development of new elements of science talk. Away from debate, where the pressure to comment on and more importantly defend the scientific legitimacy of one idea or the pseudo-scientific pedigree of another was much less, invocations of science were often cursory or appeared without any explicit identification. Implicit suggestions about the nature of science—for instance, the sorts of authorities quoted on scientific matters or the kinds of evidence marshaled to prove a particular point—certainly deserve attention. But, without knowing how people really talked about the scientific out in the open, such suggestions remain incomplete at best and an exercise in historical mind reading at worst.

Science on the Edge

As I indicated above, I have chosen to focus on four topics that elicited especially intense and revealing debate over what science was and was not: phrenology, evolution, relativity, and UFOs. These four offer a chronological sequence over two centuries crucially important in the creation of modern science-related discourse and help to make important trends in that discourse clear. Each also had a prominent public profile, attracting widespread attention in general-audience literature, including magazines, and from diverse groups and individuals. This popularity, and the numerous sources it helped to produce, makes these debates ideal places from which to judge the circulation of science-defining tools. Just as importantly, it made them contemporary sources of science talk, bringing comments on the nature of the scientific to those beyond the thick of debate as they read about the news of the day. In that sense, the controversies generated by these topics did not just reflect trends in talk about science but also helped to spread them.

Phrenology, which linked the size and structure of "organs" in the brain to the details of personality, garnered enormous amounts of attention from a wide variety of Americans in the years leading up to the Civil War. Its believers and critics also spread the rhetorical resources for making sense of science alongside those intended for making sense of one's own personality. Phrenology appeared in the stories of Nathaniel Hawthorne and Mark Twain and in the diaries of lesser-known Americans who took their sons or daughters to have their skulls examined.[52] From the 1830s on, itinerant phrenological lecturers crisscrossed the county, conducting such exams and distributing cheap books and periodicals, such as the *American Phrenological Journal*. At the same time, phrenology was surrounded by intense controversy over the scientific nature of its claims, the methods of its founders, and the relation of its ideas to accepted religion. Because it threatened to reduce the mind to the operation of the brain, phrenology straddled one of the most significant issues in contemporary science talk, namely, the relationship between natural and supernatural accounts of the world. It thus invited evaluations of the ways in which science and religion differed or, more typically, harmonized. But phrenology also opened up questions about the ownership of science and its methods. The period witnessed the very earliest stages of attempts to organize and formalize scientific practice and communication and the proliferation of often opposing desires to spread democracy from politics into all areas of life, including science.

Phrenology remained active as a source of self-help into the latter decades of the 1800s, even into the early twentieth century, but not as an arena of public controversy. It seemed that by the 1870s, all eyes had turned to Charles Darwin's evolution. It is easy enough to overstate the importance of late-nineteenth-century evolutionary debate. It did not do much to promote the trend toward naturalism that was already taking hold among many researchers or cause widespread abandonment of traditional religion. But it did throw developing patterns in scientific rhetoric into stark relief, and thus change the ways Americans generally talked about science. Evolution straddled the same fault line between the natural and supernatural as phrenology had a generation earlier. Some historians have claimed that the naturalism inherent in phrenological science helped pave the way for evolutionary accounts of life. But through the intense debate over Darwin's ideas, we can see how many Americans had come to make sense of naturalism in different ways than their forebears and how evolution prompted creative responses to traditional problems. Controversy over evolution also cast light onto a wide range of newly emerging tensions in the ways Americans discussed scientific practitioners, their methods, and their place in the world. In particular, it demonstrated more aggressive attempts than earlier in the century to distinguish science from other areas of knowledge and practice, often in the hands aspiring professionals.

Though the debate over evolution quieted in the years straddling 1900, it flared up again during the antievolution crusade of the 1920s. By that time, one of the most seemingly unlikely turns in the history of American popular science was underway, namely, the explosion in public discussion of Albert Einstein's theory of relativity. Relativistic physics began to provide fodder for headlines, news columns, and a succession of popularizations. And it became the subject of debate, not least because of its reputation for sheer strangeness. Its aura of something outside of normal understanding or typical scientific rhetoric pressed numerous observers to defend the ways they talked about science or to find new ways of expressing themselves that challenged traditional formulas. The popular discussion of Einstein's work foreshadowed the slow shift in American popular culture from depictions of biology as the archetype of proper science to one in which physics became exemplar of the scientific. Widespread attention to relativity also played midwife to the most enduring and repeated image of the scientist in the modern world, namely, Einstein himself. Ultimately, the image of Einstein the superhuman genius whose contact with mundane life was tenuous at best was one among many aspects of early-twentieth-century science talk that depicted science as unapproachable, mysterious, and beyond the realm of normal humans.

After mid-century, there was a continuing tendency for Americans to talk about the nature of the scientific most intensely where orthodox conceptions of science seemed ready to break down. That was especially clear in discussion of a number of self-proclaimed scientific topics that emerged during the 1950s and 1960s, including the unorthodox cosmological theories of Immanuel Velikovsky or the possibilities of extrasensory perception. Sometimes devotees of these fields explicitly mentioned the prior oddity of the new physics as a precedent for their own views. In general, during the latter half of the twentieth century, the new physics has been a particular point of contact between conventional and unorthodox ideas, from Buddhist interpretations of quantum mechanics to antigravity enthusiasts' interest in general relativity. First among such unorthodoxies, in time if not in attention, were unidentified flying objects or UFOs. The earliest encounter with what would quickly be called a "flying saucer" occurred on the heels of World War II in 1947. Over the next several decades, UFOs and their presumed alien pilots appeared in nearly every popular culture venue available. The desire of some investigators to incorporate UFOs into orthodox science, the willingness of others to turn away from the scientific or toward a self-consciously "alternative" science, and the horror of critics eager to protect true science from any blemish opened the doors to an orgy of science talk in popular books and magazines.

It may seem at first glance that we could and perhaps should separate these four episodes into two involving science and two involving pseudo-science. Despite the naturalness of that move, I think to do so would be a serious mischaracterization. If we mean by pseudo-science those ideas that are now dismissed by the bulk of the scientific community, it would be much more appropriate to say that each episode contained scientific and pseudo-scientific elements. Brain localization is still very much a viable idea. Alternately, Darwin's original exposition of evolution included, even as a major element, the use and disuse of acquired characteristics, a notion now roundly rejected. Special relativity has held up well over time, but many other parts of Einstein's work that I will deal with have not, particularly the details of his Unified Field Theory. Alternately, if by pseudo-science we mean untruth taking on the form of science in order to deceive or the perversion of the true methods and goals of science out of ignorance, then we are engaging in modern science talk rather than trying to understand its history. During the late 1800s, one of the most common targets of the term "pseudo-science" was Darwinian evolution.

A division of episodes into science and pseudo-science will also make us miss a point crucial to the story about the boundedness of science. Instead of alternating from unorthodoxy to orthodoxy and back again, all of these controversies threatened to rupture in some way the existing scientific rhetoric with which people approached them. That is a major reason they elicited so much comment and part of their value as lenses through which to examine the changing content of American science talk. If anything, their potentials to transgress increased over time. And that is precisely what we should expect if the boundedness of science also grew during the same period. There was no orthodox science during the early 1800s of which phrenology could run afoul, despite the fact that it has done so retroactively. Nor could the science talk of the day make such an orthodoxy. Despite the fact that phrenological doctrine carried naturalization into the most intimate depths of the human mind, the rhetoric that both its advocates and critics used did little to differentiate between science and religion or science and not-science more generally. Later debate over evolution was far more intense and divisive not simply because of new arguments for purely natural accounts of the world, but also because diffuse and bounded depictions of science often clashed dramatically as partisans sought to eject their opponents from true science.

By the 1920s and 1930s, there were an increasing number of ways to portray something as not science. But its separateness raised pressing questions about how to bridge increasingly impermeable boundaries lest science become disconnected. The seemingly inexplicable fervor with which advocates of relativity sought to make Einstein's ideas relevant, in some cases to return to a depiction of science as an integral part of the Western intellectual tradition, suggested the importance of this mission. But, as we will see, these bridge-building attempts were defeated by a science talk that had grown too good at making distinctions. Ultimately, popular discussion of relativity only reflected and helped to intensify the otherness of science. By the latter half of the twentieth century, some observers came to admire science for its otherness and to laud it for its supposed conflict with accepted wisdom. At other times, the boundaries of science seemed so formidable that, rather than scale them, those who wanted to turn science-ward began to talk about their own sciences, whether focused on UFOs or psychic powers, with rhetoric that emphasized openness and interconnection with the rest of human knowledge. For most other Americans, science was safe to ignore. This is substantially the story I will begin to tell in the next chapter.

Chapter 1   Phrenology

A Science for Everyone

In 1839, NELSON SIZER set out to help phrenologize the nation. He had been reading about phrenology on his own for years when a neighbor invited him to join in giving a public lecture on the subject. From that first experience, he was hooked. In his memoirs, Sizer recalled the "solemn dignity" of the lecture and the way his more experienced neighbor "tried to bear himself as if he were preaching a sermon." For the next forty years, Sizer devoted himself to advancing the cause of phrenological doctrine. He became a traveling lecturer, wandering from New England to Virginia in search of heads to examine and minds to enlighten. There were perils to travel in an age before highways or public transit. Once, he had to walk twelve miles, braving flood waters, in order to reach the site of a lecture. But, for Sizer, such troubles seemed worthwhile. Four decades later he still remembered the "'live men'" at the end of his travels who were "eager to be taught by anyone who can teach on any good subject."[1] Though Sizer moved to New York and took up a more sedentary practice after mid-century, his early years seemed to confirm the judgment of another phrenological enthusiast that "all naturalists are *peripatetics*."[2]

Phrenological notions also traveled far and wide in antebellum America. Charles Dickens reported that after visiting Washington, DC, on a tour of the new nation in 1842, he was occasionally asked by Americans living elsewhere whether he was impressed by the "*heads* of the lawmakers" there, "meaning not their chiefs and leaders, but literally their individual and personal heads, whereon their hair grew, and whereby the phrenological character of each legislator was expressed."[3] The center of American phrenology was not the nation's political capital, however, but its financial one. The New York City offices of arch-phrenologist Orson Fowler; his brother, Lorenzo; sister, Charlotte; and wife, Lydia; housed the largest mail-order firm in the city and drew as many visitors as P. T. Barnum's nearby museum.[4] Like Sizer and a host of other itinerant speakers, the Fowlers also took their phrenological busts and inexpensive publications on the road. Not everyone was enamored of phrenology, though. Its doctrines met derision as "quackery" or were denounced as "materialism." Skeptics accused its wandering practitioners of trying to get rich through trickery and deceit.

Though phrenology has not survived the passage of time quite so well, it was a serious contender for scientific status in its antebellum heyday. Phrenologists almost universally claimed that their chosen field was a legitimate science and thus worthy of respect and attention. In 1835, a friend of phrenological ideas in the Richmond, Virginia–based *Southern Literary Messenger* notified readers that phrenology "is no longer to be laughed at. It *is* no longer laughed at by men of common understanding. It has assumed the majesty of a science."[5] Other admirers elevated phrenology to the pinnacle of mental, moral, or intellectual science, made it the basis of a comprehensive science of man, or simply asserted its deserved place among the branches of "unbending science."[6] The Fowlers especially trumpeted phrenology's claim to be scientific to their readers and listeners, asserting its right to the label was no less than that of mathematics, astronomy, or natural history.[7] When accused of being more devoted to money than to truth, Orson Fowler exclaimed that he worked so hard because "*I loved the science!*"[8] Critics widely recognized the claims for phrenology's scientific basis even as they disputed them. To detractors, phrenology was a "pretended science," "miscalled science," or "science falsely so called."[9] Debate among phrenologists and antiphrenologists, and between advocates of different phrenological schools, frequently involved claims about what counted as truly scientific.

Like the country they crossed, however, the science that phrenologists and their critics invoked covered a huge area. Almost anything could have been called scientific and frequently was. Many common exemplars of antebellum science would undoubtedly be familiar to present-day observers. Astronomy was one routinely cited by authors of popular phrenological literature, with chemistry and geology not-so-distant seconds.[10] Indeed, chemical applications to "the great subject of scientific agriculture" were a matter of widespread discussion and concern.[11] The popular sciences of botany and natural history also occupied a great deal of attention. But other examples of science did not correspond well with modern ideas. Some people still used science and philosophy as interchangeable terms.[12] Many Americans also identified theology, music, or art as branches of science. Such usage recalled the historical roots of science as organized knowledge about whatever subject. Alongside phrenology, the Fowlers promoted a system of shorthand called phonography as scientific because it was based on specific rules. Science could even be a synonym for knowledge itself. Cultivated authors sometimes referred to a well-informed person as one of "profound science" or someone who knew little as possessing "superficial science."[13] Nor was the scientific easily narrowed down by reference to what it was not. Critics routinely joined their denunciations of phrenology as false science with attacks on its moral and religious foundations and implications. Scientific falsehoods were damnable, but carried no special attributes. Error was error by whatever standard used to measure it. Alternately, the cultural value of antebellum science was not generally to be found in the contrasts between its revelations and those of religion, art, or common sense; scientific knowledge was not important because it was different. Instead science was significant because it harmonized with truth of all kinds.

Antebellum science talk reflected the fuzziness of science as a category. Much more during the early 1800s than since, Americans talked about the "sciences" rather than "science," as if the sum total of scientific knowledge were too amorphous to make sense of all at once. There were few rhetorical means of clarifying where a unified science ended and other things began. Science was not regularly associated with a special group of experts whose pronouncements could serve as arbiters of scientific boundaries. The methods commonly ascribed to science were those of right thinking in any area, rather than uniquely scientific. It was possible to contrast science with religion to some degree, but the impulse to harmonize natural and divine truths tended to obscure scientific boundaries rather then strengthen them. Finally, the history of scientific ideas and their discoverers was the same as the histories told about truths of all kinds in whatever realms of human endeavor. All knowledge, as Sizer's "live men" might have agreed, was equally valuable and ultimately harmonized into one grand system of truth. In short, the boundaries constructed by antebellum science talk, unlike the details of the physical landscape, allowed wandering with ease. Phrenologists talked about crossing the divide between informal reading and public lecturing without much concern. Likewise, they passed unfazed over the boundary between scientific speaking and religious sermonizing. Such travels were not unique. When it came to scientific rhetoric, almost all Americans could be peripatetics.

Phrenologizing the Nation

Phrenology was a European import by way of Scotland. Friend and foe alike traced its roots back to the work of Franz Josef Gall, an Austrian physician born in 1758. As the oft-told story went, while still a boy Gall noticed that his fellow students with an aptitude for language had prominent eyes. By the 1790s, through a program of examinations, interviews, and dissections of brains, he had formalized his notion that each faculty of personality was connected to a particular organ in the brain. The larger the organ as revealed by the shape of the skull, the more prominent the kind of behavior or talent with which it was associated. The organ of language, for instance, sat imme-diately behind the eyes, a location that explained Gall's initial observations. Gall began to lecture on his new system, gaining a notable disciple in the person of fellow physician Johann Gaspar Spurzheim. In 1802, the Austrian government forbade the teaching of phrenological ideas as materialistic and therefore damaging to religion and morality. Shortly afterwards, Gall and Spurzheim left Vienna. They lectured their way across central Europe, reach-ing Paris in 1807. Here Gall stayed until his death in 1828. Spurzheim, who possessed a greater flair for public relations than his mentor, continued to spread phrenology. He found his most receptive audience in Scotland, and in 1815 began a campaign on behalf of phrenological doctrine in the *Edinburgh Review*, visiting the city on several occasions to give lectures and demonstrate the dissection of human brains.[14]

Locating the operation of the mind in the physical details of the brain appealed especially to younger Scottish physicians, many with radical politi-cal and social ties, who felt excluded by the alliance between the Anglican Church and the British state. This alliance sanctioned the notions of the rul-ing classes with divine approval, doubly tainting all political disagreement with social and religious violation. It also barred non-Anglicans from pres-tigious Oxford and Cambridge Universities. A purely materialistic phrenol-ogy challenged the spiritual authority of the church and, just as it focused on the physical brain to the exclusion of an immaterial mind, sought to divorce secular power from the mythological trappings of religion. In the tumultuous intellectual world of Edinburgh, a materialistic interpretation of phrenological doctrine eventually mixed with other radical beliefs, including Continental speculation on the nonsupernatural evolution of life. The 1844 *Vestiges of the Natural History of Creation*, an anonymously written book that sparked serious controversy on both sides of the Atlantic, proposed a purely natural account of the history of the earth from its formation to the present day. Its actual author, Edinburgh publisher Robert Chambers, was also enamored of phrenology, as were many other "transmutationists."[15]

Among the most important of Spurzheim's Scottish converts was the Edin-burgh barrister George Combe. In 1828, he published one of the most widely read and intensely controversial phrenological books of the time, *The Constitution of*

Man. Combe sought to apply the phrenological linkage of personality and behavior with physical conditions and the health of the body to all aspects of human life, particularly morality. However, though Combe sometimes used the secular implications of phrenology, he also aspired to make phrenological doctrine more respectable and mainstream than it was likely to become in the hands of materialistic radicals. He was concerned with the organization and institutional bases of phrenology as well as its intellectual spread. He became the primary force behind the formation of the Edinburgh Phrenological Society and the society's quarterly, the *Phrenological Journal,* begun in 1823. Not all of Edinburgh eagerly accepted his phrenological ideas and their widening application. He failed in his bid for the chair in logic at the University of Edinburgh in 1836, losing the post to rival mental philosopher and antiphrenologist William Hamilton. This clash represented only one skirmish in a protracted battle between supporters and critics throughout the early 1800s.[16]

Phrenology reached American shores by the 1820s in its Combean form, applied widely to social issues in a gentlemanly and learned mold. Publications, including Combe's *Constitution of Man,* crossed the Atlantic at first. Eventually phrenological ideas arrived with a number of Americans who encountered them while traveling in Europe. Charles Caldwell, a North Carolina physician and founder of Transylvania University in Kentucky, attended a lecture on phrenology in Paris in 1821 that inspired him to write the first original American volume on the subject. But it was through Spurzheim, who arrived on an ill-fated lecture tour of New England in August 1832, that phrenology took deepest root in the United States. He visited Yale, where professor of chemistry and natural history Benjamin Silliman, one of the period's most notable of American men of science, claimed that the faculty was "in love with him."[17] In Boston, he gave a series of lectures to large audiences and mingled with the city's elite.

In November, Spurzheim succumbed to a strenuous lecture schedule and a New England winter. In his final moments, he reportedly expressed no fear of death, only the desire "to live as long as I can for the good of the science."[18] His unfortunate demise, and the drama of his sacrifice, made phrenology more visible and compelling to many Americans than it had been before. Prominent citizens of Boston rose to "take up the fallen standard." The phrenological society they created included 144 official members, many from the medical profession, and inspired the formation of other phrenological societies in a number of major American cities and a host of small towns.[19] The Boston society began publication of a monthly journal, *The Annals of Phrenology,* which reprinted numerous articles from the *Edinburgh Phrenological Journal* and arranged for the republication of the works of Gall, Spurzheim, and Combe. Its members expressed a keen desire to apply the phrenological truths that had been dropped at their feet. In 1834, the society formed a committee to visit schools in the Boston area and evaluate their treatment of children according to phrenological and physiological principles.[20]

The willingness of so many Americans to take up phrenology reflected the highly amorphous nature of antebellum American science. As Nelson Sizer's introduction to phrenology suggests, scientific education and practice were largely informal and unregulated. There were no graduate training programs or other scientific accreditation organizations. Meanwhile, the circulation of scientific knowledge was not very separate from the flow of public information more generally. The contents of science were not policed by specialized publications devoted to original research or even the general cultivation of science. The lack of these things is partly what historian Alison Winter meant when she wrote that there was no preexisting scientific orthodoxy during the early 1800s.[21] She was discussing Britain but her point applies as well to the United States. Those aspiring to make scientific claims on either side of the Atlantic could not depend on the support of an orthodox establishment. The individual success of particular ideas required the case-by-case marshalling of compelling evidence, persuasive arguments, and networks of allies. This state of affairs tended to generate a great deal of talk about science, but very little means of enforcing strong boundaries between the scientific and other forms of knowledge or practice.

Some Anglo-Americans had begun to make moves toward a more formal and organized scientific enterprise, one that sought to make clearer distinctions between what was inside and what was outside science. In Britain, the Cambridge don William Whewell stood at the center of a circle of "gentlemen of science" eager to extend their power and influence over the scientific world. The early years of the British Association for the Advancement of Science, founded in 1832, marked an important period of development in their plans.[22] In America, the so-called "scientific Lazzaroni" also labored to organize, formalize, and impose their own set of standards on a scientific community, to elevate the value of science in the minds of the public, to forge their desire to study nature into a means of making a living, and to be able to pursue their work without interference from other groups or professions. Named for the *lazzaroni* who begged on Italian streets, they included director of the United States Coast Survey Alexander Dallas Bache, eventual head of the Smithsonian Institution Joseph Henry, and Harvard professors Benjamin Pierce, Louis Agassiz, and Benjamin Gould. Like their British counterparts, the Lazzaroni sought to control membership in the scientific community through the manipulation of scientific bodies such as the Smithsonian, the Coast Survey, and the American Association for the Advancement of Science, founded in 1847 on the model of its British sibling. Bache once declared that "while science is without organization, it is without power."[23]

Nevertheless, actual scientific work was diffused widely throughout American society among "farmers, tradesmen, clerical workers, and manual laborers" as much as professors and government-sponsored researchers.[24] Botany, for instance, provided a field of widespread interest and activity among otherwise ordinary Americans. Some collected plants for entertainment, others for enlightenment, and still others as a means of admiring the handiwork of God. Modern observers might be inclined to dismiss these part-time workers as "hobbyists" or "amateurs," but there was little contemporary basis for doing so. Informal botanists often had a cooperative rather than a deferential relationship with specialists such as Harvard University's Asa Gray.[25] In fact, amateur practice was one of the major routes to becoming engaged in more formal scientific activity. Lazzaroni member Joseph Henry was originally a watchmaker's apprentice who became interested in science after some private reading on the subject. His subsequent scientific education was largely self-directed. Even among those who had achieved formal status in some way, scientific practice was frequently done on the side. Faculty at the many small colleges and medical schools around the nation had to expend their own time and financial resources to conduct independent research, if they were among the minority inclined to do so at all.[26]

Likewise, there were few established sources of specialized scientific information. *The American Journal of Science,* edited by Benjamin Silliman, provided articles focused on scientific work. Throughout the early 1800s, however, the circulation of the *Journal* remained below a thousand.[27] The vast bulk of antebellum science was available either through informal channels or was mixed in with the general flow of news and information. Some families brought science into their parlors in the form of widely available textbooks and demonstration equipment. Many lecturers, either traveling from town to town or visiting established venues such as the Lowell Institute in Boston, spoke about such topics as physiology, chemistry, and of course phrenology. Astronomy was particularly popular in the wake of several comets that appeared in antebellum skies. Science also occupied an integral role in the general coursework required at many colleges. Finally, science had a prominent profile in the forming mass media of the period. At least some newspapers included a wide array of articles about natural science, from stories on the natural history of animals and plants to interesting geological discoveries. Many magazines also included large numbers of stories about scientific topics.[28]

Like scientific information more generally, American phrenology soon burst out of its initial, somewhat eighteenth-century gentlemanly form as the

1830s progressed. Local societies composed of "erudite" citizens often faltered after the first blush of enthusiasm, as did the one formed in Cincinnati in 1831. Frances Trollope, one of a number of Europeans to visit the new nation and record their impressions, noted that by the third meeting of the Cincinnati society, the one at which dues were to be collected, only the treasurer was in attendance. Even the anointed Boston society eventually lost momentum. Its *Annals* expired after only two years, partly for want of subscribers.[29] By the time Combe arrived in America to lecture in 1838, the heart of phrenology increasingly beat within a highly commercialized and popularized version of the science championed most visibly by the Fowler family. The Fowlers aimed not at the learned, but at ordinary people through frequent lecture tours and inexpensive popular literature. In his later years Mark Twain recalled Orson Fowler's visits to his boyhood home of Hannibal, Missouri, where his lectures and head examinations were "popular and always welcomed" and where people began to adopt phrenological terms themselves, batting them "back and forth in conversation with deep satisfaction."[30] Unlike the *Annals*, the Fowlers' more entertaining and eclectic *American Phrenological Journal*, founded in 1838, survived into the early twentieth century, where it became *Life* magazine. By the end of the 1840s, the *Journal* had one of the largest readerships in America, and its audience verged on the truly national, numbering in the tens of thousands by the Fowlers' estimates, and including subscribers from the East Coast to the wilds of Wisconsin. The "scribbling Fowlers" also published a mountain of inexpensive books on phrenological and other subjects, meant to be accessible to the general reader.[31]

The Fowlers' version of phrenology represented a larger attempt to "Americanize whatever in science and the arts, is capable of improving or adorning the mind, or otherwise benefiting mankind."[32] Phrenological doctrine was intended to be the harbinger of a new and better world, but it was not alone. In their typically exuberant typescript, the Fowlers exclaimed that "Reform, *Reform*, REFORM, is emphatically the watchword of the age."[33] In addition to phrenology, the Fowlers championed dress reform, claiming that the narrow waists of dresses did damage to women's bodies; hydropathy, which saw the cure to all kinds of disease in the application of water; and the building of octagonal houses, which not only used space better but improved health and happiness. Their New York offices became a meeting place for Americans interested in all kinds of contemporary causes, from abolitionism to the crusade against alcohol. Lydia and Charlotte Fowler were both active in campaigns on behalf of women's rights and made their New York offices a gathering place for like-minded reformers. Lydia received the second M.D. degree awarded to a woman in the United States and became the first to hold a professorship in medicine.[34]

The examination of heads occupied a central place in Americanized, re-
form-minded phrenology. It was, as the Fowlers frequently claimed, one of the
bases on which phrenology could be considered a "demonstrative science."[35]
Whether out of a desire for entertainment or a faith in the power of phrenological prediction, a diverse collection of prominent Americans offered their
heads for examination: newspaperman Horace Greeley, future president James
A. Garfield, abolitionist John Brown, the Mormon prophet Joseph Smith,
poet Walt Whitman, the famous Siamese twins Chang and Eng, and the Spir-
itualist Andrew Jackson Davis. Some, such as Clara Barton and Ulysses S.
Grant, reportedly chose their careers based on their craniological attributes.
As more and more Americans filtered into cities, leaving behind the familiar
occupations of their fathers and mothers and in the wake of hard economic
times in the late 1830s, professional advice may have led many parents and
young people into the New York offices of the Fowlers or the rented rooms of
traveling lecturers. The head examination became a staple of the itinerant
phrenologists and one major reason for their popularity. Lectures were often
free, but exams, and a booklet in which one's attributes could be marked, cost
a modest fee.[36]

The Fowlers' version of phrenology appealed to so many Americans because it harmonized with the growth of democratic sentiment, particularly
during the second quarter of the nineteenth century. Andrew Jackson's occupation of the White House coincided with widespread celebration of the common man. There was a general blossoming of new -ologies, -isms, and -pathies
intended for use by nonspecialists and on whose behalf adherents claimed the
mantle of science.[37] In medicine, for instance, systems that brought knowledge
and the ability to use it to the typical person, such as Thomsonianism, eclecti-
cism, and hydropathy, flourished. The growing sense that individuals were able
to do whatever they wanted for themselves was not limited to science. The
forces of democratization reshaped Christian attitudes and practices over the
early part of the nineteenth century, as Americans responded to the tension
among competing denominations by making the individual conscience a more
important arbiter in discussion of faith.[38] This opened the door to a variety of
new religious groups from Mormons to Millerites that questioned established
religious orthodoxy. Others, most notably Robert Dale Owen and those who
joined his utopian community at New Harmony, sought innovation in the social realm. Meanwhile, formal barriers to the scope of individual initiative fell.
The number of states with licensing laws that privileged certain practitioners
in law and medicine declined precipitously between 1800 and 1860.[39]

Visions of an informed and competent citizenry also provided powerful
motivations for diffusing scientific knowledge among the masses rather then

allowing it to be controlled by an intellectual aristocracy. Slogans such as "knowledge is power" found endless repetition among inhabitants of the young republic, harmonizing with the image of the intelligent and informed citizen as the engine of democracy. Another common saying of the period declared that "knowledge, like money, depends on its circulation."[40] As new, democratically charged sources of information exploded in the 1820s and after, the spread of science reached unprecedented levels. American newspapers, and the science articles they carried, multiplied. Cheap editions of books, many about scientific topics, began to appear on both sides of the Atlantic. Advances in steam trains and the growth of the postal system, which experienced a threefold expansion from 1820 to 1840, helped to distribute this literature. New organizations, such as public libraries and political parties, further prompted the spread of all kinds of information. There was also a proliferation of public lectures. During the 1820s and 1830s, "public oratory entered a 'golden age,' becoming a vehicle for all sorts of reformers and educators, politicians and salesmen."[41]

The glee with which many antebellum Americans violated social and intellectual boundaries, including the crowds that reportedly threw a rowdy party on the White House lawn to celebrate Jackson's election, was not universally praised. The Lazzaroni and others seeking to formalize and organize science did not welcome the uncontrolled diffusion of scientific practice and knowledge either. It was, in fact, one of the major motivations for their efforts. Both Bache and Henry were concerned about "quacks" distorting true science. Likewise, those devoted to a more strictly bounded science began to resent the stress on popularization. Bringing scientific knowledge to the masses sometimes became a barrier to generating original research meant for one's peers. At the very least it was a distraction.[42] Though attempts to establish a scientific orthodoxy were not entirely successful during the early part of the nineteenth century, the 1830s and 1840s were crossroads in the formation of separate "elite" and "popular" forums in the Anglo-American world.[43]

In particular, despite its success in reaching ordinary people, the practice of reading character by reading heads, or "bumpology," as it was sometimes dismissively called, was the target of special distaste. Those dedicated to a more learned, generally Combean model of phrenology often found the practice somewhat vulgar. Though some articles in the *Annals of Phrenology* seemed implicitly to condone assessing personality through observation of the skull, at least one member of the Boston Phrenological Society denounced it for turning "a dignified science into a system of legerdemain." The author urged the cessation of head exams as a means "to aid in promoting Phrenology in a way more in accordance with scientific taste." In the published recollections of his visit to the United States, George Combe himself made the oft-noted comparison between reading heads and reading palms and lamented that the practice "is said to have created a strong feeling of disgust against phrenology itself in the minds of men of science and education." Even the revered Spurzheim was enlisted in the battle against head reading. An obituary notice in the *American Journal of Science*, edited by Silliman, noted that the great man had "refused to designate the characters of living individuals by the application of the rules of his science." The Fowlers, however, disputed this claim.[44]

Critics outside of phrenological circles likewise ridiculed the practice of examining heads. But the materialistic tendencies of phrenology were far more worrisome to many of them. Enoch Pond, a professor at the Bangor Seminary, claimed that phrenology was turning the "noble science of mind" into "'a mere Golgatha—a place of skulls.'"[45] These tendencies were sometimes linked to previous eruptions of scientific naturalism—that is, accounts of the world that pushed aside divine power and influence in favor of explanations based on purely material phenomena. Phrenology followed on the heels of debate about the nebular hypothesis, which claimed to explain the formation of the solar system from a primordial cloud, and geological work that questioned such standards as the role of Noah's flood and the age of the earth.[46] According to some of its detractors, a naturalistic view of science reached its purest and most unacceptable form in the positive philosophy of the French philosopher August Comte, whose works began attracting attention in the English-speaking world during the early decades of the century. Comte and his followers, opponents charged, cared only for what could be measured and tabulated, privileging mathematics, astronomy, and physics and dismissing the claims of metaphysics or theology to scientific status, much to the ire of many religious thinkers.[47]

Most American antiphrenologists relied on the work and opinions of Europeans such as William Hamilton and Charles Bell to defend their positions. But, by the late 1830s, they could cite several homegrown authorities as well. *An Examination of Phrenology*, an 1837 work by Washington, DC, physician Thomas Sewell, exerted considerable influence among critics. Sewell opposed phrenology primarily on anatomical grounds, claiming that the varying thickness of the skull and the frontal sinuses made links between head shape and personality meaningless. As for the brain, Sewell asserted, anatomical examination further invalidated phrenological claims for multiple organs. David Reese's *Humbugs of New-York*, which appeared the year after, attacked phrenology on broader ground, lumping it in with a variety of excessive errors, such as mesmerism and abolitionism. And although opponents of phrenology were perhaps less quick than traveling lecturers to resort to the new sources of information that flooded antebellum America and used them less skillfully, their objections did appear in periodical articles, including attacks in the *North American Review* and *Christian Examiner*, and were thrown down from the lecture podium. Sewell's book was based on lectures he gave in Washington, DC, in 1828.[48]

Science and the People

Despite the efforts of critics such as Sewell, the lack of any clear social or intellectual scientific orthodoxy made it difficult to expel phrenology from the realm of science once and for all. Likewise, the specifics of antebellum science talk did more to preserve the breadth of science as a category than to erect strong scientific boundaries. It was much easier to claim something was legitimate science than to justify why it was not. Americans, for instance, even the most determined critics, never claimed phrenology was unscientific because it was not done or believed in by "scientists." Nor did its supporters call phrenology a science based on scientists' approval. The absence of the term "scientist" was due partly to its relative novelty. William Whewell coined the word in 1834. Benjamin Gould independently suggested it fifteen years later. There were other existing terms for people who cultivated science, such as "man of science" or "scientific man," but they were used in highly informal ways without distinguishing between specialists and dabblers. They did little to evoke the image of a unified and organized scientific community. Whewell's generation of a new label for scientific practitioners in particular was ultimately motivated by what he perceived as the weakness of antebellum descriptors and a desire for a new comprehensive category for scientific workers as a definite class.[49]

The relative weakness of Americans' sense of a clear-cut scientific community was reflected in phrenological rhetoric. Few Americans discussing phrenology made reference or appealed to such community, either in general or by citing its most prominent members. The science that most phrenologists talked about was not the possession of any particular group of people, either men of science or scientists. Anyone could do it and therefore anyone could legitimately comment on what belonged inside and outside of its boundaries. Such contemporary devotees of the study of nature, such as Benjamin Silliman, Edward Hitchcock, William Herschel, or Alexander von Humboldt, all of whom have frequently become central figures in historical narratives, went almost unmentioned.[50] Neither their judgments nor their examples were used widely to fix the boundaries of the scientific. The contemporary figures whose heads offered exemplars of prodigious or admirable development were almost entirely drawn from a field that may have had much more active relevance and cachet for citizens of the new republic: politics. Among the examples and

authorities that American phrenologists cited most frequently, Daniel Webster came first, followed by Henry Clay, John Quincy Adams, and Martin Van
Buren. Such well-known politicians may have occupied a status not unlike
star athletes or singing sensations in the modern day.

   As in matters of politics or faith, individual judgment empowered by slogans such as "knowledge is power" sometimes became the ultimate determi-
nant of scientific status. An author in the *New England Magazine* informed
readers that, in the battle over the truth of phrenology, "the arena is the land
of freedom; the arbiters, the American public."[51] Phrenologist after phrenologist urged their readers or listeners, as members of an "enlightened and reflecting public," not simply to accept what they were told, but to observe and
decide for themselves about the truth of phrenological principles.[52] The Fowlers were perhaps the most enthusiastic purveyors of democratic rhetoric. The
introductory statement of the *American Phrenological Journal* declared that in
America, unlike Britain, "knowledge is here every man's birth-right; and a science whose tendencies are to elevate its votaries to the greatest heights, or to
initiate them into the deepest mysteries is alike, the property of *all* our citizens,
who have the inclination and ability to investigate and acquire it."[53]

   Appeals to the common man allowed phrenologists to dismiss some of
their critics' scientific judgments. A correspondent in the *American Phrenological Journal* claimed that while common people accepted phrenology, those who
have "pinned their faith on the sleeve of the learned professional" were still
blind to its truth.[54] Meanwhile, some of those same critics disputed depictions
of a science that was equally the possession of everyone. Many antiphrenologists offered a far more elitist view of knowledge in which the masses were simply unreliable. Thomas Sewell noted that ordinary people had always wanted
to discover the secrets of nature without "tedious observation, and laborious
research," an illusion encouraged by phrenology. Others mourned the sad fact
that authors "must give a high flavor, a quick relish, a powerful zest, to whatever
is intended for American consumption," so that, "although a shrewd and intelligent people, our political, intellectual, and religious condition, is one continuous subjection to quackery and humbug."[55] Scientific practitioners who aspired to organize their profession and to promote original research rather than
popularization also sometimes took a dim view of the public. In such circles, the
word "popularizer" itself began to take on a negative tone.[56]

   Elitist depictions of science were rare. Alternately, invocations of the wisdom of the general public did a very poor job of justifying any kind of scientific
expertise independent of common sense, and this was a distinction that even
the most passionate champions of "a science *for the people*" sometimes felt the
need to make. "Was not the *term* science appropriated to [astronomy, geology

and the natural sciences] severally . . . long before the great mass of people believed in their truth, or even heard of their existence?" asked the Fowlers.[57] By loosening the strong connection they made between science and popular knowledge elsewhere, they could make space for their own phrenological expertise. Indeed, the Fowlers' business relied to a certain extent on their position as phrenological experts. They rarely articulated any awareness of this situation, but now and then they did admit to looking forward to the establishment of phrenology as a "profession" along the lines of medicine or the law.[58] Andrew Boardman, an American admirer of Combe, was much more explicit about his own desire to construct a well-bounded phrenological practice. In a book defending phrenology he devoted the last third to a discussion intended for "professional men," which suggested that knowledge did not belong to everyone equally.[59]

However, even when Americans wanted to distinguish between those who could legitimately pronounce on the truth of phrenology and those who were mere "pseudo-scientific men," they had few rhetorical means of doing so.[60] A very small number of Americans of any kind debating phrenology invoked education, credentials, or degrees as sources of authority. Institutional affiliation was mentioned now and then. Institutions and organizations were increasingly important landmarks on the national landscape. An 1844 article on the impacts of phrenology claimed that "much of the vice and misery of mankind grow out of our institutions, so many of which clash with the nature of man."[61] Institutions could also be repositories of legitimacy. Many of the testimonials included by Boardman in his defense of phrenology came from men with connections to universities, colleges, or asylums. There were some moves toward a more organized American phrenology, too. In 1849, the Fowlers hosted a convention to establish an American Phrenological Society. Conventions, they noted, were "the order of the day."[62] But the sanction of the society almost never appeared in the Fowlers' bids for scientific legitimacy. Nor did the majority of phrenologists and antiphrenologists stress membership in certain key institutions or organizations, either for themselves or those they cited.

Instead, many phrenologists confronted critics on the grounds that they simply had not devoted enough attention to the study of the science, perhaps a natural move considering the lingering link between science and general knowledge. When Thomas Sewell offered the opinions of John Quincy Adams and Daniel Webster as evidence against the truth of phrenology, Boardman protested that such supposed authorities had hardly taken up the study of phrenological principles in any serious way. But no observer sought to make clear how much study was enough, or how a person was to evaluate the learning of

someone else. In the absence of defining credentials, the ability to pronounce
on scientific matters in phrenology often came down to personal choice. Benjamin Silliman, who attended Spurzheim's and Combe's American lectures
and certainly had more access to books and articles than most citizens, chose
not to call himself a phrenologist because he had not studied the science "*as a
whole*," while a host of itinerant lecturers such as Nelson Sizer did assume the
title, without much obvious soul-searching.[63] A reprinted lecture in the *American Phrenological Journal* asserted that anyone who believed that the brain was
the organ of the mind and recognized the different functions of the various
cerebral organs could legitimately claim the title of phrenologist.[64]

Similar kinds of rhetoric could be used to limit public participation in
phrenology, but with little more clarity. Phrenologists who took up Combe's
more hierarchically oriented phrenology rather than the Fowlers' populist version sometimes placed stronger limits on who could speak on phrenological
matters, though in ways not so different from their "practical" brethren. In
the first of his lectures in New York, Combe told his audience that he wished
"to impress on your minds that it is not by attending a course of lectures, that
you can become fully acquainted with phrenology." Nor, observed George's
physician brother, could one simply read a book or pamphlet and suddenly
become a phrenologist. The notion resembled the error of believing that one
could "become the equals of Liebig or Faraday, by reading a volume on chemistry." But little about the story of Combe's own conversion to phrenology
did anything to set the science beyond the reach of the average person. He
had studied phrenological ideas for several years, sending from Edinburgh to
London for skulls, books, and other materials.[65] Such a tale contained social
elements, suggesting that only people who had enough free time and sufficient
resources could become phrenologists. But this claim was never made explicit
in American public discussion. Any linkage of phrenological expertise with
class would have smacked heavily of aristocracy. Rather, in the United States
at least, the Fowlers' thriving mail-order business promised to bring phrenological materials to everyone. The widespread demonstration of phrenological
principles in lectures did the same.

A second major means of discerning legitimacy involved an evaluation of
character, something desirable no matter what activity one engaged in. The
fiery Charles Caldwell accused Thomas Sewell of dishonesty and phrenology's
critics generally of "piracy in science" by denying phrenologists the credit
for their discoveries. The issue of character could be turned on phrenologists themselves. The Fowlers criticized "drunken phrenologists" for spreading
"quackery, dishonesty, or infidelity." Meanwhile, they recommended the lectures
of Mr. D. G. Derby because he appeared "to be thoroughly imbued with the

spirit of Phrenology, blended with the spirit [of] philanthropy."[66] Critics of phrenology, and some friends who preferred its more aloof Combean form, almost universally stressed the pecuniary and commercial motivations of the traveling head readers, including the Fowlers and other "itinerant self-seekers" who often catered to their clients' vanity with flattering assessments of character. Those engaged in phrenological debate were not alone in reaching for character as a sign of legitimacy or quackery. During the 1840s, fellow Lazaronni Joseph Henry and Alexander Dallas Bache both stressed the character of true men of science.[67] But though true science depended on and could help develop a lofty character, few Americans claimed that moral rectitude was the sole possession of scientific devotees or that aspiring to be scientific was in any way more important than seeking to be righteous.

Bacon and Beyond

Just as science was often depicted as a communal possession, there was little about the ways its methods were described that set scientific knowledge apart as special in any way. No one amidst debate over phrenology made any reference to a "scientific method." In general, antebellum Americans rarely used this term. Instead, there were a number of other important, much less specific words around which discussion of methodology turned. The fact was among the most important of these.[68] Enthusiasm for facts often came at the expense of theory and hypothesis, words Anglo-Americans, phrenological or otherwise, rarely used in a positive sense. Advocates declared that phrenology regarded a fact "as worth a million" theories, and its practitioners took "great care . . . to avoid everything hypothetical." Others trumpeted the claim that phrenology had been "demonstrated CHIEFLY BY A WORLD OF PHYSICAL FACTS."[69] In one of his American addresses, George Combe was so anxious to relieve Gall of the charge of hypothesizing, he noted that "in the disjointed items of information which he first presented to the public, there appears a want of ordinary regard for systematic arrangement." Rather, a "candid and uncolored statement of the facts was all he seemed desirous of furnishing."[70] For Combe and many American phrenologists, conclusions that did not limit themselves to facts did not deserve the name of science.[71]

Once these facts were in hand, they guided the process of discovery, generally through their classification. Some phrenologists claimed to be led by "a thousand or more facts converging to" the truth or by letting "the facts *classify themselves*," that is, "*wholly by induction*."[72] The truth would emerge as patterns and coincidences became clear. The invocation of facts and induction provided a common ground between those on all sides of the debate over phrenology. In an American edition of an antiphrenological work, Paul M. Roget, professor of physiology at the Royal Institution, asked, "Who will dare to set up his opinion [against] ascertained facts?" The central issue for Roget and other skeptics concerned the "reality of these facts on which so much is made to depend" and the legitimacy of phrenologists' inductions. Such critics accused phrenologists of "hasty generalization," supposition, and a priori reasoning when "facts must be accumulated for ages" before conclusions were warranted in other areas of science. Those who disputed the "extremely hypothetical" conclusion of phrenologists, alternately, employed "masterly induction."[73]

Talk about facts and induction sometimes culminated in appeals to the method of Baconianism. Phrenologists of all kinds noted that their "favorite science" had been "perfected, by the true Baconian method of inductive philosophy," rested securely "on the canons of the Baconian philosophy," or was "strictly Baconian in all its parts and processes."[74] The inspiration for Baconianism was the late-sixteenth- and early-seventeenth-century English essayist, courtier, and natural philosopher Francis Bacon. Bacon had argued that the proper means of finding truth proceeded inductively from "senses and particulars, rising by a gradual and unbroken ascent, so that it arrives at the most general axioms of all." As the seventeenth century ended and the eighteenth progressed, Anglo-American authors increasingly cast the study of nature in Baconian terms. By the early 1800s, Bacon had achieved a kind of iconic status. In 1845, a columnist in the Southern Quarterly Review announced that "Lord Bacon, of all philosophers, was the only true one," who divined the only "natural and equitable mode of acquiring knowledge"—that is, proceeding inductively from particular facts to generalities.[75]

Later generations of Americans sometimes looked back on fact-centered Baconian induction as a very limited view of scientific thought. But, in fact, popular antebellum discussion of methodology was actually quite flexible. One phrenological lecturer celebrated Davy Crockett as a good Baconian because he once advised, "Be sure you're right, then go ahead."[76] Many other phrenologists recommended the use of analogy. Orson Fowler, for instance, "boldly" asserted "that all real 'analogy' is an unerring guide to truth," arguing that phrenology deserved the label Baconian because it depended on analogical reasoning. When one had learned a general principle inductively, the next step involved applying it "to all new but analogous facts."[77] The use of analogy was a critical piece of most phrenologists' arguments for the notion of plural cerebral organs and the idea that larger organs indicated a stronger propensity for certain kinds of thought or behavior. Elsewhere in nature, they noted, different functions implied different organs. Likewise, natural examples of parallels between size and use abounded; the large muscles of the blacksmith's hammering arm showed this clearly.

Phrenologists and their critics did widely condemn theory and hypothesis, but they typically used those terms only to mean ideas with absolutely no foundation in reality. Instead, phrenologists of even the most popular sort made room for general ideas that might be described as theoretical in the modern sense. A preponderance in favor of reflection led to "metaphysical *theorizing,* which is valueless." But if the observing powers dominated, then one would be unable to ascend from facts to principles, which was "the only possible means of arriving at *truth.*" Real phrenologists aimed at the "fundamental laws" and "fixed natural rules" that operated in nature. At such times, facts appeared to be little more than a means to an end. Lydia Fowler, sister of Orson and Lorenzo, echoed these sentiments in a book for children. She counseled her young readers to "think, inquire, and be not satisfied with simple *facts,* but search for the *principle,* and endeavor to understand it."[78] Putting such principles to work frequently became another yardstick of their truth. Both "observation and application," a columnist in the *Cincinnati Mirror* noted, "form the test of scientific doctrines." Sometimes such application simply meant providing an understanding of events. In an article on the "Utility of Phrenology" in the *American Phrenological Journal,* the author noted that phrenology rested on the same basis as Copernicanism—"*it explains phenomena.*"[79] At other times, the truth of phrenological doctrine was in its power to reform the habits of individuals and the society in which they lived.

Such descriptions of the methods of science gave antebellum Americans considerable flexibility, but they did little to help make science appear distinct or unique. With a little practice anyone could use Baconian induction, meaning that science was easily blurred with ordinary knowledge. Many guides to phrenology included directions for conducting head exams. Orson Fowler advised all his readers to "try the experiment, and proclaim the result."[80] Phrenology itself helped to show that every skull contained the very same organs of observation, capable of recognizing "form" and "color," and reasoning, including "comparison" and "causality." Other kinds of methodological notions were widely accessible as well. Analogy was known more generally as the "Yankee prerogative."[81] There were attempts to make methodology more exclusive and therefore susceptible to the exercise of specifically scientific expertise. In Britain, William Whewell's *Philosophy of the Inductive Sciences* (1840) defended the more aggressive use of hypothesis in science. His views were used by the leadership of the British Association for the Advancement of Science to bar provincial members from "Baconian *participation*" and to make them into either "the minions of or the deferential *audience* for the Association's theorists."[82] Nevertheless, Whewell's attempts had no noticeable impact on American discussion of phrenology's scientific status. Whewell was very rarely mentioned by either supporters or opponents of phrenology.[83]

The widespread availability of facts in every department of life made it difficult to confine the idea of science in many other ways. There was some precedent for singling out certain kinds of facts as deserving of special attention. Phrenologists universally rejected the results of introspection as valid sources of information about the mind and its operation. That was the method of "metaphysicians," who indulged in flights of imagination. Defenders of more traditional introspective observation, frequently citing the work of British philosopher John Stuart Mill, responded that the products of such observation were themselves legitimate pieces of data. What was more, any true study of the mind needed to begin with such facts. Devotees of Combean phrenology attempted to limit the pool of legitimate facts even more stringently. In his American lectures, Combe commended the use of "the most striking cases as best calculated to bring the truth to light." Similarly, George H. Calvert, in a compendium of articles from the *Edinburgh Phrenological Journal,* claimed that original discovery came only through the most "extreme cases."[84] The effect of such a methodological twist was to depict a science potentially removed from the real world of most Americans.

But in their choice of legitimate facts, many Americans returned to broad and weakly bounded visions of science. The majority of American phrenologists stressed the particular importance and value of those facts found in everyday living. Thus anatomical dissections weighed little in their estimation. Though they "secured the respect of learned men," dissections neither attracted nor were generally open to ordinary people without access to instruments, detailed training, and human brains other than their own. Like the introspection of the metaphysicians, which took place "in their own closets," this lack of accessibility made anatomy's value as a means of testing scientific truth suspect. Instead, the truth of phrenology was a matter of "daily and constant *personal* experience" for the phrenologists and their audiences. Head-reading thus became more than a source of advice; it was the guarantor of phrenology's truth. "The truth is," proclaimed a correspondent to the *American Phrenological Journal,* "a man can no more become a good phrenologist without traveling and handling heads than he can a mineralogist without traveling and handling minerals."[85]

Like the facts on which it was based, the methods that antebellum Americans talked about guaranteed the correctness of scientific knowledge, but they were not uniquely scientific in any way. None of them did much to distinguish science, to make it or its methods special, or to easily reject given ideas as unscientific. Reformed Protestants relied on inductive rhetoric in discussions of biblical interpretation and the quest for solid religious principles. Artists and poets also invoked facts and induction in the search for the proper, realistic representation of the world. Nor did phrenologists, eager to protect the scientific claims of their own field, dispute such broad application. Rather than turn to a well-known man of science for an example of a large organ of comparison, the basis of the inductive method, the Fowlers pointed instead to William Shakespeare.[86] Non-fact-centered rhetorical tools did little more to separate science from nonscience. A search for principles and general laws applied to nearly every area of thought, from cooking to law. Analogy could draw out important conclusions from any field of study. Nor was the test of application the sole province of science. In the "practical age" of the early 1800s, truth of every kind was widely expected to make some real impact on the world.

Antebellum talk about method ultimately provided science with few clear boundaries. Areas of confused or unsystematized ideas were not science, though they were perhaps simply awaiting organization through scientific generalizations. It often seemed that only a few things clearly remained outside the potentially scientific fold: pure fantasy or notions that had no application in the broadest sense. Both of these were more easily summed up as simply wrong than as especially unscientific. Nor did any of the most common methodological notions have the slightest verbal link to science, a rhetorical fact that seemed to emphasize their general place in good, rather than particularly scientific, thinking. Later generations' common invocation of "scientific method" often amounted to little more than warmed-over induction. But their appeals did call on science more clearly and more directly than any piece of early-nineteenth-century methodological vocabulary. Ultimately, this lack of a rhetorical link to science left the scientific status of fields that used similar modes of thought unclear. The fact that religious truths could be classified and reduced to general principles led at least one antebellum observer to assert that Christianity was "as much a science as mathematics."[87] Another author similarly suggested the formation of a science of sculpture that would provide rules for artists who wanted to create realistic images.[88] Yet it was also possible to claim that religion or art used inductive principles or consisted of systematized knowledge without mentioning anything about science at all.

An Eminently Religious Philosophy

Despite a few claims otherwise during the early 1800s, religion was one of the most frequent examples of a valid area of knowledge that was not science. For instance, some antebellum Americans made reference to "science and religion" as categories separate enough to require a conjunction. But, while comparison of science with religion generally or Christianity in particular offered opportunities to make science distinct and unique, they were also among the most powerful means of breaching scientific boundaries. A very few Anglo-Americans promoted a deep and radical division between scientific and religious realms. For most other observers, a potential boundary between science and religion became the occasion for rhetorical bridge building and reconciliation rather than divorce. Efforts to show how scientific and religious truths were harmonious, how they always supported one another, and how they cooperated seamlessly in revealing a consistent picture of creation all but erased the borderline that prompted attempts at harmonization in the first place. Linking scientific knowledge into a more general network of truths about the world often obscured anything special about science. Such a harmonized science was significant, but largely because of its close ties with knowledge of the divine rather than because of any independent scientific value.

Because of concerns about the irreligious and atheistic implications of explaining the mind in terms of the brain, phrenological debate often started Americans talking about the relationship of science and religion. Any division between them was not based, as it sometimes was during the last third of the 1800s, on ascribing scientific knowledge to men of science and formal religious knowledge to theologians. Nor was there much methodological distinction between the two. Rather, they were most typically distinguished by their subject matter. Science began with those worldly facts that were accessible to human observation. Religion began with sources of information that were beyond the reach of unaided human reason. For Christians, these otherwise out-of-reach facts came though the divine revelation recorded in the Bible. In practice, however, no such absolute division was easy to make. Almost all branches of early-nineteenth-century Christianity, Protestant ones in particular, looked for evidence of divine power in the natural world as well as in the pages of scripture.[89] The human mind was an especially important area of overlap between the scientific and religious, natural and divine. It was the means by which people experienced and rationally understood the world around them. It was also the recipient of divine inspiration and the subject of divine influence during a conversion to Christianity. In even suggesting that there was no such thing as an immaterial mind, phrenology could be and sometimes was charged with denying the existence of everything from the human soul to God.

Though science and religion were not distinguished by separate methods, there was a methodological means of avoiding any potential conflict, what some modern scholars have called the Baconian compromise.[90] This compromise both assigned to science whatever could be observed and reasoned out by human faculties and set those facts derived from divine revelation beyond question. The association of Bacon with the attempt to limit science to its proper sphere added yet another layer of meaning to invocations of his name. An 1839 article asserted that since phrenology was "strictly Baconian in all its parts and processes, it pushes its investigations no farther than the safe ground of observation and experience, and pretends only to examine the proper subjects of rational enquiry."[91] Safe ground came to an especially abrupt end on the question of the mind. The majority of American phrenologists steered well clear of statements about its ultimate reality or role. Mariano Soler, professor of modern languages at the College of Louisiana, informed a lecture audience that the essence of the mind and soul lay beyond the reach of phrenologists.[92] Another phrenological lecturer asserted that phrenology, though offering powerful insights into the manifestation of the mind, did not speculate about its true nature.[93]

Some supporters of phrenology found the distinction between science and religion that lurked in talk about the human mind very useful. This was even the case among those without radical social or political ties. George Combe regularly differentiated between science and religion in ways that heightened the value of the former, at least in some contexts. In particular, Combe juxtaposed the two categories when he was advocating a system of secular education linked to natural science rather than one steeped in religious dogma and involving tales about the supernatural.[94] Nor was he especially concerned with avoiding any conflict between these two realms. One phrenologist criticized his treatment of religion by noting that Bacon's name never appeared in the *Constitution of Man*.[95] However, Combe's fairly strong division between science and religion was rare among American phrenologists, even if it allowed them to proclaim the value and power of science. If anything, their response to the possibility of conflict between the scientific and religion was nearly the opposite of Combe's. Rather than suggest that phrenologists replace religion with science, a correspondent in an 1853 issue of the *American Phrenological Journal* asserted that phrenology needed lecturers who were both steeped in the science and "devout men," with phrenology "in one hand and religion in the other."[96] Others also depicted phrenology not only as a science but a "soul-elevating science."[97]

Many American devotees of phrenology depicted their science as perfectly harmonious with religion on a wide variety of levels. Caleb Ticknor claimed that phrenology came to the same conclusions as Christianity, but was condemned for it.[98] A variety of other phrenologists did much more than simply avoid making statements about the human mind. They used phrenological doctrine to show that a nonmaterial aspect of the world existed. Phrenologists commonly argued that the existence of the cerebral organ of spirituality, which was not coincidentally located at the top of the head, "is proof that man has an immaterial nature, a spiritual existence. . . . It is demonstrative proof. It is the highest possible order of proof. It settles the matter completely."[99] By the latter 1840s, some phrenologists had even struck an alliance with mesmerism to form the new field of "phreno-mesmerism" as a means of confirming the reality of and exploring the world beyond purely physical things. Despite frequent charges of materialism and fatalism, few lists of the uses of phrenology failed to mention its value to religion, both in showing the natural rules by which God had intended humans to live and in offering support for our spiritual nature. Phrenological analysis also promised to determine the precise nature, and therefore affirm the reality, of religious conversion and the changes it could make in the human brain.

In general, despite or rather because of the sensitive nature of discussing the relationship between mind and brain, phrenologists in the United States turned to religious language to talk about supposedly scientific matters with remarkable ease. Many drew from the still very vibrant rhetorical reservoir of natural theology. William Paley's *Natural Theology* (1802) was "among the most popular textbooks on any subject for academy and college students in antebellum America" and a commonly used resource in such widespread pursuits as botany.[100] Natural theological statements and concerns circulated widely among the opponents of phrenology and among many phrenologists too, particularly the Fowlers. Authors sometimes attributed the success of inductive reasoning to the coherent and single order behind all phenomena in all areas of experience ensured by God's careful design of the universe. The use of analogy relied on a similar faith in the unity of nature in all its forms. Likewise, according to some phrenologists, all truths were useful because they came from a wise and benevolent creator.[101] Through such methods, phrenology revealed "the hand-writing of God." Its principles, like all truth, bore the "unmistakable evidence" of "divine origin." The Fowlers included in their repertoire of lectures one on natural theology, and Orson Fowler wrote in his treatise on memory that instructors should not teach "children ANY thing in science or nature, without teaching the GOD in it all."[102]

Sometimes phrenology itself took on a religious air. In many testimonials, authors noted their initial resistance to phrenological truths, a resistance overcome in most cases only by long study and laboriously gathering facts for themselves. In other accounts, or even in the same ones that stressed hard work and observation, becoming convinced of the correctness of phrenological doctrines took on the appearance of evangelical religious conversion. These stories helped both to blur the line between natural and religious truth and to emphasize the importance of their recognition by individuals rather than experts or institutions. Occasionally, such conversions occurred in the wake of a near miraculous reading of character from a cranial exam or an encounter with Gall or Spurzheim. Robert MacNish, a Glasgow physician whose account found its way into several American works, recalled greeting phrenology with ridicule. Later, after a private meeting with Gall, a "new light dawned" in him and he finally saw the truth. The cover of the *Annals of Phrenology* included an engraving of Gall's profile with beams of light radiating from his forehead. Other converts compared the advent of phrenology itself to a "revelation from heaven to pour light on the world of mind" or credited Gall and Spurzheim with dispelling the dark clouds around mental philosophy and allowing the truth to shine forth "in all its simplicity and irresistible splendor." In the wake of their work, "we all bowed, and acknowledged the truth."[103]

The strong mixture of phrenology and mostly evangelical Protestant Christianity peddled by many American phrenologists sometimes met resistance, but not in ways that helped to clarify the distinction between science and religion. An 1847 article in the *Methodist Quarterly Review* blasted the Fowlers in particular for judging the truth of religious ideas by comparison with phrenological doctrine.[104] Such denunciations also suggested that the Fowlers had violated some boundary line, but their transgressions were often framed with respect to developments internal to early-nineteenth-century Christianity. Charges that phrenologists were trying to substitute the "'religion of nature' for that of the Bible" tapped into fear of sectarianism and the formation of such groups as Millerites, Seventh-Day Adventists, and Mormons as much or more than the illegitimate mixture of religion and science or the domination of the former by the latter.[105] David Reese attacked phrenology not because it threatened to overthrow Christian religion but because it represented one of the many new "sects in science, philosophy, religion" that had begun to "overrun our beloved country."[106]

On both sides of debate over phrenology, the prominence of religion sometimes obscured a clear view of science. This tendency extended beyond phrenological discussion too. While antebellum Americans made reference to science and religion in general, in specific instances they often framed issues in terms of geology and religion, astronomy and religion, or phrenology and religion. Phrenologists did perhaps use more religious language than devotees of other sciences. Such rhetoric made potentially dangerous ideas safe. Phrenological doctrine also confronted prescriptions of behavior and morality, topics that were without doubt on religious ground. Lorenzo Fowler went so far as to depict phrenologists as completing the work of Jesus in reforming human behavior.[107] But the choice to make science religious rather than to make religion scientific, at least rhetorically, was still telling. Phrenological maps of the skull themselves echoed a more concrete sense of religion than of science. There was no single organ related to science alone, though there were several to which many phrenologists ascribed special religious significance, such as "veneration" or "spirituality." This definiteness reflected the greater social establishment of Protestant Christianity during the 1800s and made talk about religion a valuable resource. Lecturers sometimes relied on the social structures of religion to do their work. As Sizer recalled, they often spoke in churches, which were frequently the center of many antebellum communities, often with the aid of the preacher.[108] Likewise, many phrenologists seemed more than happy to put religious vocabulary to work communicating their ideas and showing why they were important.

A Common History

Stories about the history of science helped to reinforce the strong links between it and other areas of knowledge, including religion, in ways that further obscured what was unique about scientific practice or knowledge. Historical accounts showed nothing special or distinctive about the development of science or about its famous devotees. Instead, its progress followed a course common to all kinds of truth about the world. Many phrenologists certainly connected phrenology's past with its rightness. The Fowlers declared that the "history of the DISCOVERY of phrenology, furnished ample demonstration of its truth," and Gall frequently joined the ranks of methodological saints. Comte reputedly called him the Francis Bacon of mental science, a dubious honor in the eyes of the religiously orthodox. Opponents of phrenology attacked Gall for beginning not with observations but with a preconception—that is, the multiplicity of organs in the brain—and only observed to confirm his "favorite hypothesis." Gall's many experiences successfully predicting character in schools, asylums, and prisons could also be used to depict non-fact-centered views of method. In her children's book on phrenology, Lydia Fowler noted that as a young person, Gall spent a long time contemplating the harmony and order of the universe in addition to particular facts. Even those who disputed the truth of phrenology sometimes suggested that Gall be given his due in accomplishing what Isaac Newton did for heliocentrism, that is, proving true what had been simply conjectures about the localization of brain functions.[109]

Gall was not the only historical figure to appear amidst debate over phrenological ideas. Antebellum Americans betrayed little awareness of a contemporary scientific community, but history provided a wide selection of deceased men of science among which to locate exemplars. Beside Bacon and Gall many public commentators mentioned Newton, both as a model of the genius and in the context of his work on natural laws. In his *Constitution of Man*, Combe used the law of gravitation as an archetype for all other laws, physical, organic, and moral. Other phrenologists stressed that the "laws of life and health are as immutable as the pillars of heaven, and can no more be violated with impunity, than the law of gravity."[110] Some authors compared Gall's connection of the sizes of his classmates' eyes and their proficiency with language to Newton's experience with the apple, in both cases using the accidental nature of the discovery to deflect any charges of preconception or unseemly attachment to hypothesis. Newton's name often became a kind of benchmark. One phrenological advocate claimed that even if only one of Gall's cerebral organs actually existed in nature, Gall would outshine Newton in importance. Likewise, just as Bacon often came to symbolize fact-based induction, analogy was sometimes linked with the name of the French naturalist Baron Georges Cuvier, who was said to be able to infer the whole structure of an animal from a single bone.[111] Galileo Galilei and William Harvey also found frequent mention as important men of science, as did John Locke. Finally, Benjamin Franklin occasionally represented the American contingent of past giants of science.[112]

The oft-cited trio of Newton, Galileo, and Harvey held particular importance because their individual stories often illustrated a larger tale about the history and progress of scientific knowledge generally, a tale that also showed how the search for such knowledge paralleled the search for truth in other realms. Phrenological writers widely noted that contemporaries at first resisted the discoveries of Newton and company. Such rejection had been a fate shared by Robert Fulton, Christopher Columbus, Edward Jenner, and Michael Servetus. Benjamin Silliman asserted in his *American Journal of Science* that every important science first met with rejection. To supporters, nonscriptural geology had proved Silliman true, as had phrenology, which labored under charges of materialism and fatalism. The identified source of such resistance varied among authors. For some, the orthodox were simply sluggish in seeing new truths. For others, typically those who saw themselves as upholders of tradition, the overeager cultivators of young sciences misapplied their conclusions or overextended themselves through speculation, only to calm down with time. Whatever the cause of conflict, almost everyone who addressed this topic noted that geology had passed through its "long and vexatious trial" and was now harmonious with other areas of knowledge, mainly Christianity, as phrenology would someday be. That golden day had not yet arrived, however, and phrenologist after phrenologist pointed to the persecution of their science as the latest in a long list of times when truth had been rejected. So frequently did they make such claims that critics ridiculed their invocations of Copernicus, Harvey, and even Jesus Christ.[113]

The use of geology as an example or the re-telling of the story of Galileo may suggest to a modern observer that contemporaries had in mind a struggle of progressive science against conservative religion, but this was not the case. During the 1830s and 1840s, the Galileo story almost never implied anything about the relationship between science and religion. Instead, Galileo's ideas were rejected by all sorts of people simply because they were new and unfamiliar.[114] The initial rejection of truth found one of its more compelling examples not in the history of science but in the history of Christianity. Both the story of Jesus Christ and the early years of the Christian religion in the Roman Empire seemed to suggest the resiliency of truths in the face of suffering, perhaps even that a greater degree of suffering implied a greater truth; hence the presence of Jesus among the likes of Copernicus and Harvey. Gall's ejection from Austria and flight westward under threat of persecution called to mind for some the experiences of the apostles of the New Testament. Nelson Sizer noted that as "it was with the religion of Jesus, so it was with the apostles of the true mental philosophy." The sacrifice of Spurzheim, whose public lectures contained moral lessons that "would not have dishonored the Sermon on the Mount," made such comparisons all the more powerful.[115]

The Utility of Science

On the whole, most of antebellum American science talk did little to give science the appearance of a unique and exclusive category. There was no special class of persons who were widely cited as arbiters in scientific matters. Science was something anyone could, and sometimes should do. Depictions of science as a communal possession helped to keep the lines blurred between scientific work and political, religious, and even commercial enterprise, where participation and circulation were as significant as production. During the second quarter of the 1800s especially, all these endeavors were frequently given a democratic spin. Hence, for many Americans, popularizing science was as important to the validity of scientific knowledge as original research. Nor did the methods of science make it any more distinctive. Appeal to fact, induction, Baconianism, analogy, higher principles, and application were all presented as parts of right, rather than especially scientific, thinking. Such approaches may have given scientific information its stamp of legitimacy, but did little to set it apart as different from or more reliable than anything else. Indeed, Americans affirmed again and again that all truths from all areas of experience, including those gained from nature and revelation, needed to be harmonious. Looking back in time, many observers saw further confirmation that all real knowledge of the world shared a common history.

Antebellum scientific rhetoric constructed images of science that were thus accessible on many levels, paving the way for Americans to wander unhindered through what historian Susan Faye Cannon called the early-nineteenth-century "truth-complex," that grand scheme of knowledge into which separate statements about the world were expected to fit together smoothly.[116] Contemporaries could openly talk about the shared conclusions of science and religion or the shared methods of scientific and artistic work. The particular points they made may have been disputed by others but there was nothing inherently transgressive about describing science in the same terms as other kinds of knowledge. By the same token, false notions, including those about nature, were denounced often as humbugs, quackery, or charlatanism. Such terms applied equally to claimed errors in the moral, religious, political, and even commercial realms, from abolitionism among supporters of slavery to P. T. Barnum among supporters of sobriety. For present-day observers, however, the amorphous boundaries of antebellum science have occasionally presented a challenge. They have nearly begged for the imposition of modern definitions to find a sense of solidity amidst the often vague rhetoric. Depending on their perspectives, some historians have come to radically different conclusions regarding the cultural authority of science during the period, characterizing it as both the "norm of truth" and "relatively insecure."[117]

The early-nineteenth-century place and prestige of science was ambiguous and perhaps even paradoxical, as we might expect, but it was ambiguous in a very different way than during the twentieth century. During much of the 1900s, as I outlined in the last chapter, the power of science originated in what made it different and unique. For many of its admirers, it was valuable because it was so unlike other, less reliable sorts of knowledge. The inaccessibility of its content in some ways spoke for its rightness. To its critics, or those simply looking for a reason to think about other things, these same attributes warranted rejection or dismissal. By contrast, according to historian Thomas Broman, eighteenth-century scientific pronouncements gained their authority from the appearance of accessibility and openness. Those who made them spoke "for anyone sufficiently apprised of the facts to formulate a scientific comprehension of the matter."[118] The nineteenth century was a transition period between these two modes of influence. During the early 1800s especially, science often gained its prestige not by being separate from other kinds of knowledge, but by harmonizing with them.

Certainly, a science with few clear boundaries, and with conceivable connections to almost every area of human experience, possessed a wide relevance to the world of all Americans, at least potentially. Some subsequent observers have depicted the second quarter of the nineteenth century as a period of obsession with practicality and usefulness, implying that Americans only valued scientific knowledge for its economic or material benefits and that practitioners of science seeking public support felt a new pressure "to *persuade* others of its utility," which "had not previously been so strongly felt." The need to show some material benefit from their work supposedly prompted some men of science to make extravagant and even irresponsible claims.[119] A purely material view of utility may have reigned inside the halls of legislative bodies, limited by constitutional mandates and often minimalist views of government, but more popular notions of practicality were marked by nothing so much as breadth. Insofar as science was a potent word in the early 1800s, it was at least partly because of its continuity, or even identity, with the rhetorical, intellectual, and social influence of religion, for instance. Controversy over the usefulness of phrenology occurred, an 1839 article in the *American Journal of Phrenology* noted, not because there was any doubt on this point, but because there were so many different views of utility.

Amidst the debate over phrenology, it was typically the breadth of science that made it important. Utility could mean producing some tangible product, such as a machine or wealth, but knowledge could also be practical through the "moral, physical, and intellectual elevation of society" or simply by enriching the knower.[120] The first of phrenology's many uses listed in *The Mirror of Nature* (1839) by the Bostonian, self-described "practical phrenologist" John Fletcher was its value "in *itself* as a *matter of fact.*" Nowhere was the practical value of phrenology more highlighted than in the works of the Fowlers, including the *American Phrenological Journal.* And yet, an 1840 article in the *Journal* probably written by Orson criticized the "poor and pitiful" view of those who judged science by "its results in railroads, patent churns, and improved window glass, and which does not place first and foremost 'the forming of the soul of man' and its high culture, as that is secured and advanced by the progress of science." Only later, particularly in the years around mid-century, did notions of application come to focus on the making of machines. By 1860, a list of dates worth remembering in the *American Phrenological Journal* consisted almost exclusively of those associated with some technological innovation, such as the invention of stereotyping, airguns, and the thermometer.[121]

Just as in the twentieth century, however, there was another side to claims about the importance of the scientific. It was hard to ignore a science whose potential influence seemed to have such enormous scope. But that

same diffuseness sometimes made science as a category somewhat more difficult to work with. Cognitive scientist George Lakoff claimed that humans
tend to prefer using what he calls mid-level categories—those that correspond to typically encountered and easily identified groups, such as dogs
or trees, rather than those that appear large and abstract.[122] Antebellum
Americans demonstrated such a preference in a variety of ways. Though
there were many articles in contemporary magazines about scientific topics
or mentioning science in their texts, only a small fraction of these invoked
science in their titles. During the next century, even to some extent later in
the nineteenth century, when science was a more concrete term, this dis-
parity was smaller.[123] In such cases, authors and editors were not using science as a means to classify or summarize stories or to attract readers' initial
attention. A similar tendency occurred in Americans' talk about sciences
as opposed to science and in the tendency to frame talk about natural and
divine truths in terms of religion and specific areas of science, whether astronomy, geology, or phrenology, rather than science itself. An amorphous
category such as science may not have had the intellectual impact that these
more specific branches of knowledge possessed. Likewise, discipline-specific
labels such as astronomer, chemist, geologist, or physician were used more
frequently in a number of American magazines than "man of science."[124]
For all their concern that phrenology be labeled a science, the Fowlers, and
most phrenologists generally, never claimed to be men of science. Neither
did most of their critics.

During the early 1800s, the role of science as an important keyword in
American culture was still in the making. There were some voices, particularly
those aspiring towards a professional scientific practice, that were advocating a
more well-bounded and independent science. Andrew Boardman questioned
phrenology's harmony with all other truths as a legitimate test of its correctness. Though he granted that such harmony made phrenological principles
more probable, he wanted to judge the doctrines of phrenology on their own
terms.[125] He was clearly in the minority among Americans debating phrenology, but similar demands would become louder as the century progressed.
Americans would increasingly talk about science as something unlike other
kinds of knowledge, distinguished by its unique practitioners and methods. As
later generations discovered, to make science truly powerful and potent as an
invocation required establishing much stronger boundaries around it, though
ultimately at the cost of making it inaccessible and potentially remote.

Though phrenology itself continued to be a source of talk about science into the late 1800s, it was not as much a focus of attention as earlier in
the century. The postbellum years brought a mixed fortune to the American

phrenological enterprise. A new generation of Fowlers, including Lorenzo's daughter Jessie and half-brother Edward, rose to prominence during the 1860s and 1870s. And when Charlotte Fowler's husband Samuel Wells died in 1875, Charlotte continued efforts to modernize and expand the firm, becoming its president when it was finally incorporated in 1884. Phrenology itself spread to the West. In 1874, an unidentified phrenologist, possibly Orson Fowler, visited Dodge City, Kansas, where he was invited by a delegation led by Bat Masterson to examine the heads of various thieves and cattle rustlers.[126] But phrenological science, suspended somewhere among the commercial, evangelical, and scientific interests of the day became somewhat marginalized in the newly organized and increasingly bounded world of the late nineteenth century. In 1863, Lorenzo Fowler, perhaps sensing this trend, abandoned New York for London with the intention of bringing American-style phrenology to Britain.[127]

By the late 1800s, some former phrenologists had moved from an interest in phreno-mesmerism to one in psychical research, or drifted toward discussions of geology and evolution.[128] Like the later phrenologists, numerous Americans looked away from phrenology and toward evolutionary ideas to explain and manage the direction of social change they witnessed around them. In various hands, evolution could reduce complex social and economic processes involving individuals, groups, races, and even nations to comprehensible terms. It could also be used to chart the course of both distant past and uncertain future, and to fix the limits of the physical world. And it helped to make visible the continuing debate over the nature of science.

Chapter 2   Evolution

Struggling over Science

In an 1872 issue of a magazine called *Old and New*, the Reverend George Axford recounted his discussions with a former schoolteacher named Mary Alden. Alden was struggling with reconciling her religious convictions and recent scientific developments, particularly evolution. Axford, himself interested in the latest science, sought to comfort her through a variety of means. He explained that religious faith had been purified by each new discovery about nature; yet there remained as much real evidence for the existence of God as for the existence of light. He also compared the seemingly deterministic elements of Darwinian evolution to those of Calvinism. But Alden would not be so easily soothed. She lamented that "everybody knows that scientific men are knocking our churches down about our ears; and to me there is a dreadful fascination in their theories. Are they made of sterner stuff than we or do they really suffer as I do when I look upon my dead friends, and wonder if this is the last of them?" Her husband could not understand why scientific men bothered her so much. But she felt she could not avoid their ideas, which appeared "in every newspaper and circulating library." Sadly, she admitted, "I can't help reading Huxley and Tyndall."[1]

Not everyone reacted quite so strongly as Alden, but no scientific topic after mid-century generated as much attention in America as evolution. Evolutionary ideas, one observer claimed, were "in the air" like "electricity, the cholera-germ, woman's rights, the great mining boom, and the Eastern question."[2] One reviewer of Charles Darwin's work claimed that the English naturalist's voyage on the *Beagle* possessed the epoch-making significance of Christopher Columbus's

transatlantic trip, of particular interest to many Americans in the years leading up to 1892. Another enthusiast compared the importance of 1859, when
The Origin of Species appeared, with 1543, when Copernicus published his
claim that the earth revolved around the sun.[3] For those more interested in
looking forward than backward, Darwin's ideas "opened a new chapter in the
history of the universe."[4] Mary Alden was not alone, however, in feeling that
all this attention was a little too much. The Canadian naturalist John William
Dawson complained in 1872 that "we now constantly see reference made to
these theories" of evolution "as if they were established principles, applicable
without question to the explanation of observed facts" even in the pages of
"popular articles" and "text-books."[5]

Talk about evolution was accompanied by an enormous amount of talk
about science. An evolutionary history of life potentially challenged a wide
variety of antebellum assumptions about nature. Similarly, evolutionary science sometimes seemed to clash with accepted ideas about the nature of the
scientific. Alden was fairly typical in feeling prompted by evolution to recon-
sider the relationship between science and religion. Other Americans reflected
on scientific methods in light of Darwin's achievements or felt compelled to
compare Darwin himself with an image of the ideal man of science. For some
of its most ardent admirers, evolution not only revised the fundamental truths
of biology, but altered the pursuit of scientific knowledge more generally. The
American geologist Joseph LeConte, a prolific popularizer in his own right,
asserted that the restriction of evolution to the realm of biology was a widespread fallacy. Rather, he triumphantly informed his readers, "nothing since
the birth of science (unless, indeed, we except the heliocentric theory of the
solar system) has so extended the domain of science, and so enlarged the intellectual horizon of man" as evolutionary ideas.[6] Alternately, many critics lamented that, "instead of designating what is most rigorous, exact, and assured
in human knowledge," natural science under the influence of evolutionary
ideas was "fast becoming identified with what is most fluctuating, hypothetical, and uncertain in current opinion and belief."[7]

The science that so many late-nineteenth-century Americans appealed
to, reconsidered, defined, and debated sometimes resembled the one that their
parents and grandparents talked about during the heyday of phrenology. The
scientific could still cover enormous territory, from the study of natural to
artistic to religious truths. But many other depictions of science contained
a new sense of boundedness—that is, science became more and more recognizable in contrast to what it was not. The most notable trend was, as Mary
Alden knew all too well, growing perception of a boundary dispute between
the scientific and religious. The distinction often involved narrowing science

down to purely natural explanations of the world versus religious supernaturalism. But even when particular attributes were not clearly determined, a sense of deep difference frequently remained. Alden asked Axford to tell her "how these things appear to healthy minds, which are large enough to hold both science and religion."[8] The very notion that a mind needed to be unusually large to harmonize scientific and religious truth was a new one in popular discussion. Nor was religion the only example of not-science that contemporaries could describe. Public discussion was crisscrossed with nascent distinctions between science and art, philosophy, or common knowledge. Almost no one used science as a synonym for general knowledge after mid-century. Science did frequently continue to include at least certain aspects of mechanical invention and industry, but even here there were attempts to draw lines between scientific knowledge and its application.

The unsettled nature of talk about science had an effect on Americans' vocabularies. Some people continued to frame discussion in terms of the "sciences," invoke "Baconianism," or mention "men of science." At the same time, the greater boundedness of science manifested itself in the spread of a variety of now familiar elements of science talk, such as "science and religion," "pseudoscience," "popular science," "scientist," and "scientific method," that helped to locate where science ended and something else began. By the 1870s, statements such as "science says" or "science tells us," which portrayed the definiteness and individuality of science in almost anthropomorphic terms, became a common sight in magazine articles. Other authors referred to Science, with a capital S. Such newfound distinctness made it possible to depict science as far more important than could have been done with most of the rhetorical tools of the antebellum years. But while stronger boundaries offered a means of declaring the independence and specialness of science, they also opened the possibility of confining its significance. Alden noted in one of her letters that there were two kinds of scientific men, those who said what they could about purely material matters strictly "according to the scientific method," and those arrogant and dogmatic types who claimed that science itself was the whole of truth.[9] In the end, whether one chose to reject narrowed views of the scientific or use them to fix the cultural place of science in new ways, it was clear that Americans talking about science were increasingly hemmed in by often unfamiliar boundaries that left some of them, like Mary Alden, uncertain which way to turn.

Evolutionary America

Like phrenology, the concept of biological evolution first took on a more formal and substantial form at the very end of the eighteenth century. Over the course

of the 1800s, however, the two followed increasingly divergent paths: the one becoming the basis for a popular, evangelical, and commercial movement and the
other providing a rallying point for an emerging scientific establishment. Evolution—or transmutation, as it was more commonly referred to in English during
the early nineteenth century—found its Franz Josef Gall in the person of French
naturalist Jean-Baptiste Lamarck. His notions, like those of Gall, incorporated a
fair amount of Continental materialism, as well as a large dose of Enlightenment
faith in progress.[10] Lamarck proposed an explanation of the history of life from
its spontaneous generation from nonliving matter and through its transmutation
into different species. Two mechanisms guided this transmutation. The first was
an inherent drive toward greater complexity. The second involved what came
to be routinely called the inheritance of acquired characteristics, or more simply
use and disuse, and involved passing on traits that a given individual developed
during its lifetime or allowed to whither away. The classic example of use and
disuse was the neck of the giraffe: each generation stretched its necks a little
more to reach the high leaves on trees, passing its exertions on to its children,
and over time resulting in the distinctive form of the modern animal. However,
Lamarck and his ideas were eventually marginalized by his fellow French naturalist, the younger and much more politically savvy Georges Cuvier.

By the early nineteenth century, transmutationist ideas no less than phrenological ones had become enmeshed in the social turmoil and conservative
reaction to the French Revolution, most acutely felt in Britain.[11] Eventually,
transmutation found its George Combe in the form of Edinburgh publisher
Robert Chambers, an admirer of Combean phrenology. Chambers had instructed
himself on a wide array of scientific topics and was convinced that transmutationist ideas explained the history of life, just as slow, natural development
explained the formation of the physical universe. But he was concerned about
the radical and often materialistic implications that had been associated with
such ideas. His 1844 *Vestiges of the Natural History of Creation*, which traced the
development of everything from the solar system to human beings by appeal to
the operation of purely natural laws, was an attempt to rescue transmutation
as a valid and respectable scientific notion. The developmental mechanism
Chambers proposed linked what was already a claimed parallel between the
history of life revealed by the fossil record and the stages through which an individual embryo passed on its way to its final form. In Chambers' scheme more
advanced species gestated longer. Such an account was a strongly progressive
one. In assuming an innate progressive drive, it also had a Lamarckian edge,
and Chambers' use of Lamarck increased in later editions.

Chambers filled the *Vestiges* with evidence gathered from astronomy, geology, biology, and on occasion phrenology. His detailed arguments introduced

new audiences to transmutationist ideas. Inexpensive editions of the book and its sequel, *Explanations*, sold widely. The young Alfred Russell Wallace, also an admirer of phrenology and eventually co-discoverer with Darwin of natural selection, found inspiration in Chambers' work.[12] But others were not so enthusiastic. The *Vestiges* met with heavy criticism from representatives of the Anglican institutions of church and state in Britain, such as Cambridge's William Whewell. Indeed, the *Vestiges* sparked a sensation and resulted in fiery denunciation on both sides of the Atlantic, particularly for the limited role of divine power in its account. Chambers depicted God as having set up the initial system of natural laws that guided physical and biological development. But once events were set in motion, He was little more than a spectator. This scheme threatened the natural theological enterprise of identifying the evidence and lessons of God's direct design and care in nature. Devout geologists, such as the American Edward Hitchcock, routinely linked their interpretations of geological history with biblical events. In his widely circulated *The Religion of Geology* (1851), Hitchcock defended natural theology, including miraculous interventions in earth's history, from the author of the *Vestiges* and others.[13]

Chambers certainly expected his views to cause controversy among defenders of the alliance between natural and religious knowledge. That was one reason he chose to publish the *Vestiges* anonymously. But he did not anticipate the overwhelmingly negative reaction to his work by the majority of men of science, including some of those who were later to accept Darwin's work. Such responses involved more than just an attempt to avoid a rupture between science and religion. Rather, in a few cases, they echoed the shifting social boundaries around scientific work, the growing division between scientific insiders and outsiders, and emerging claims for the right to speak for science as a whole. In addition to attacking the *Vestiges'* mistakes and flawed arguments, naturalist Thomas Henry Huxley, later "Darwin's bulldog," predicted that when its author was revealed, he would be "prominent neither in the mechanical nor any other department of even *one* science." He further worried about the impact of the "foolish fancies" of the book on the "popular mind."[14] Likewise, in a review of Chambers' response to his critics in the *Vestiges'* sequel, *Explanations*, Asa Gray complained about those "unfit and disinclined for research" who found it "easier to speculate than to examine" and who "snatch at the results obtained by others . . . and build splendid theories vastly attractive of popular admiration." A few pages later, Gray disputed Chambers' claim that specialized members of the "scientific class" had a right to judge the general merit of the *Vestiges'* account. Instead, the matter was reversed. Gray asserted that "any such man, deeply versed in a single department, is much better qualified to judge of the whole scheme, than one who, like our author, professes to possess only a superficial acquaintance with any branch of science whatever."[15]

Such reactions were a taste of things to come. Over the course of the 1800s, much of Anglo-American experience with science remained informal and unstructured. Even after mid-century, there was as yet no robust scientific establishment, no powerful arbiters of what science was and was not, no groups of people or sources of information that could claim uncontested authority to determine correct scientific practice or resolve debates over scientific knowledge. A large number of, even most scientific workers still learned to do science by teaching themselves or by studying with individual men of science they admired. The lack of clear social and intellectual boundaries continued to encourage broad and amorphous depictions of science. No studious or industrious soul was to be left out of the scientific fold, few pieces of information were necessarily out of bounds. At the same time, the more organized and formal science envisioned by the Lazzaroni and others grew increasingly real throughout the last decades of the nineteenth century. Within an emerging scientific profession, the actual doing of science, the constitution of scientific information, and its flow out into the general culture became increasingly structured, standardized, and controlled. More bounded portrayals of science often accompanied these developments; practitioners seeking to organize their field were often highly motivated to talk about science as distinct from other things in order to protect their work and the knowledge they generated from the meddling of "outsiders."[16]

Trends in education, work environment, and organization all began to reduce general access to aspects of scientific practice and the right to speak about scientific knowledge. More formal training provided one potential hallmark of a newly exclusive scientific community. Three percent of Americans employed as astronomers before 1859 possessed a Ph.D. Twenty-one percent of those working during the next four decades had one. Over the last half of the nineteenth century, more and more researchers also took up formal positions in developing institutions of higher education or the expanding government bureaucracy. Such changes in the social location of science offered a unique set of experiences and standards, around which a select group of practitioners was forming. Shifts in scientific practice also touched those outside of that select group. The relationships between academic and amateur botanists grew increasingly hierarchical in favor of the former as the century progressed.[17] Professional associations also began to make distinctions between working men of science and other members. In 1874, the leaders of the American Association for the Advancement of Science changed its constitution to restrict office holders to those with an accomplished record of research.[18]

Shifting social and intellectual boundaries around science had a profound impact on the next wave of controversy over transmutation inaugurated by Darwin's *Origin of Species*. In the *Origin*, Darwin argued both for the biological development of species in general and for a new mechanism to explain this development, what he called natural selection. Unlike the notions of Lamarck, Chambers, or other nineteenth-century transmutationists, natural selection involved no innate drive toward complexity. Instead, beings with traits that aided their survival were able to pass those traits on to their offspring more effectively than beings without such adaptive advantages simply because they were more likely to survive. However, Darwin's contribution of natural selection did very little to make transmutation any more palatable to many naturalists. Just as in the case of the *Vestiges*, some prominent men of science strenuously criticized the *Origin* in the years after its publication. In the United States, Louiz Agassiz, the widely known Swiss transplant and Harvard naturalist, became one of Darwin's most vocal opponents. By Agassiz's death in 1873, however, the vast majority of his colleagues appeared to accept biological development in general terms, though very few gave natural selection a primary role in the history of life. Nor, for that matter, did Darwin. He left open the possibility of multiple mechanisms, including Lamarckian use and disuse, in the *Origin*.

The very different fates met by the *Origin* and the *Vestiges* in the emerging scientific community were partly due to the distinctions between their authors. Even during the mid-1800s, Darwin was not the only person to be pursuing transmutationist ideas. Railroad-engineer-turned-philosopher Herbert Spencer had begun to publish his thoughts on biology and social development. Alfred Russell Wallace formulated the notion of natural selection independently while working in Southeast Asia. A letter to Darwin asking for an impression of his ideas helped to prompt the publication of the *Origin*. But, unlike these other men, Darwin had made a substantial name for himself among those who studied nature. This reputation carried some weight. Wallace, who came from a working-class background and was largely unknown within the larger naturalist community, readily deferred to the more gentlemanly Darwin in matters of significance and priority. The careful and closely argued style Darwin used in the *Origin*, which he intended for working naturalists rather than the general reader, further added a note of seriousness to evolutionary ideas that previous works such as the *Vestiges* had not. Darwin also modestly avoided the highly charged topics of the origin of life and the development of humans, again unlike earlier and more controversial works.[19]

Just as significantly, a group of young, institutionally active, and outspoken naturalists had gathered around Darwin in the years leading up to 1859 and mobilized to defend his ideas. In magazine treatment of evolution, the particular triumvirate of Darwin, Huxley, and John Tyndall found frequent repetition.[20] Darwin's champions were not in total lockstep; they had some very different ideological and scientific commitments. While Tyndall did offer occasional public support for Darwin, he was not as deeply invested in Darwin's defense as Huxley. Tyndall's comments were part of a much larger interest in a naturalistic view of the universe. Alternately, Asa Gray, in whom Darwin had confided about his evolutionist leanings two years before the *Origin*, closely aligned himself with Darwin's ideas, even natural selection. When Agassiz attacked evolutionary notions, Gray argued Darwin's case, along with several other sympathetic souls. Yet Gray, who was a devout Christian, also tried hard to make a place for God's direct influence in biological development by making Him the cause of new characteristics on which natural selection could then act. Whatever their differences, however, most of Darwin's supporters shared a vision of a well-bounded and self-consistent science that should not have needed to turn outside itself to supernatural explanations of the world. They likewise tended to advocate an autonomous scientific community that could pursue its investigations without having to defer to the judgments of nonmembers. Ideas about biological development provided valuable vehicles for accomplishing both of these goals.

Outside of the emerging scientific world, discussion of evolution also took place amidst shifting social boundaries. Initially, the *Origin* met with relative silence. A reviewer in *Debow's Review*, an American business and agricultural magazine, mentioned that the book was interesting "for those curious about natural history" and very readable, but nothing more. On both sides of the Atlantic, most large-circulation periodicals ignored Darwin's work during much of the 1860s.[21] Some Anglo-Americans clearly expected the *Origin* to quickly go the way of the *Vestiges*. But by the 1870s, public discussion of "evolution," which had largely replaced transmutation to describe developmental accounts of the history of life, exploded. In part this was due to the more explicit extension of evolutionary ideas to humankind. Darwin's *Descent of Man* (1871) was one of a number of attempts to explain human society and culture through the operation of evolution and natural selection. A literary column in *Appleton's Journal* noted that sales of the *Descent* were "almost unprecedented" for a scientific publication.[22] Meanwhile, a number of naturalists, popular authors, including Joseph LeConte, and enthusiasts began to transform evolution into the keystone of a grand view of the dynamic universe and a central narrative for everything from the lives of stars to the fate of human souls in the afterlife.[23] Spencer was, in particular, widely credited with expanding evolution from a purely biological theory to a truly cosmic system in which the "the law of Universal Evolution" guided the development of biological, physical, mental, and social processes.[24]

Such formulations of evolutionary ideas fit nicely with many late-nineteenth-century Americans' perceptions of the nature of progress in a modern, industrializing nation such as the United States. Progress was a much-celebrated idea, sometimes in utopian terms.[25] Some of its admirers continued to look to the judgment of ordinary citizens rather than experts or elites as the source of that progress. Just as earlier in the century, a variety of popular movements emerged: religious, such as Theosophy and Spiritualism; political, including the Equal Rights Party and the Grange and Farmers' Alliance; and medical, for instance, chiropractic and osteopathy. Such efforts were only enhanced by the spread of information and ideas, aided by advances in transportation and communication technology. A correspondent in the *Catholic World* waxed eloquent about the telegraph, noting that "we have linked our disjointed world by an electric flame that flashes knowledge throughout its circle instantaneously."[26] Yet flames could easily spread out of control. The Civil War provided a particularly violent example of the challenge of holding together a single democratic nation in the face of its citizens' conflicting beliefs. Democratization could become fragmentation, the loss of any reliable standard or center. Asa P. Lyon, while advising readers how to deal with the growing problem of religious skepticism, lamented that "it is an age of impatient inquiry. Men don't readily accept the results reached by their fathers."[27] Likewise, an 1875 review of Charles Darwin's work in *Scribner's* complained that people were predisposed toward revolutionary ideas, such as "free principles" in government and radical notions in history and science.[28]

Many areas of late-nineteenth-century American public discussion were driven by the tensions between democracy and control that were coming to a head in an increasingly complex and segmented nation. The war between the North and South was ultimately a victory for centralization; aside from preserving the union, the massive war efforts it involved prompted a variety of standardizations that replaced purely local or individual measures, such as a uniform gauge for railroad tracks and standard clothing sizes.[29] More general political rhetoric turned away from a focus on individual "natural rights" and toward concern with more corporate entities such as "the state" or "society."[30] After the war, social innovations such as large corporations and a truly national mass media emerged to control and direct the course of economic activity and the spread of new ideas, though sometimes inspiring hostile reactions of their own. The last quarter of the 1800s was also peppered with often violent labor strikes and protests against the perceived corruption and excess among powerful elites.[31] The move of a number of occupational groups to segment knowledge into manageable chunks under the control of special groups, professionalizing men of science among them, represented a particularly powerful effort to impose order on the world of information. The creation of orthodoxies of knowledge eventually resulted in a kind of fragmentation that itself hastened the bifurcation of American culture between elite and popular forms.[32] But a scientific establishment also became one basis for visions of a strongly bounded, newly unified, and expert-controlled science that could provide a means to tame an unruly society and make way for the wonders of progress.

The popular diffusion of scientific knowledge also showed signs of greater organization and standardization. Some contemporaries depicted the 1870s and 1880s as a boom time for public interest in scientific works, frequently those involving evolutionary ideas. One columnist noted in 1873 that "the 'general reader,' who only a few years since had recourse for amusement to the last new novel, almost exclusively, is now much more deeply interested in the last new view of Darwin, Tyndall, Huxley," or other prominent figures.[33] The general presence of mass media articles about science and of those mentioning science showed evidence of increasing during the 1870s after what may have been a slight mid-century lull.[34] But scientific knowledge was also slowly becoming concentrated in certain locations on the information landscape. These included a proliferation of specialized journals focusing on particular aspects of scientific work, such as that in agriculture or astronomy, and sometimes aimed at communities of emerging experts rather than at general readers.[35] Science-focused magazines for larger audiences also appeared. The most notable was *Popular Science Monthly*, founded in 1872 by Edward L. Youmans. Youmans had edited *Appleton's Journal*, a publication he had conceived of primarily as a means of spreading the latest scientific ideas. When his intentions were frustrated, he moved on to found the *Monthly*.[36] Even periodicals that retained a more eclectic and newsy perspective, such as the very popular *Harper's*, sometimes segmented scientific information in special sections or columns.[37]

Instead of flowing more or less freely through the same conduits as other kinds of information, a small but growing amount of scientific knowledge was grouped into a genre of its own. The growth of popular science as a class of literature during the late 1800s did not necessarily represent a renewed commitment to making scientific ideas available to the masses, already a frequently repeated imperative earlier in the century. In some quarters, the opposite was true. For those aspiring to form an autonomous community of scientific experts, the generation of new and increasingly specialized knowledge intended for peers often overshadowed communicating this knowledge to the general populace.[38] Rather, much of the popularization of science during the 1870s, 1880s, and 1890s was about containing and managing the diffusion of science and tended toward the creation of a hierarchy of communication, from producers to consumers, that mirrored the one slowly emerging in scientific practice. Many periodical articles and books, such as the "Half-Hour" series, including *Half-Hour Recreations in Popular Science* (1874) and *Half-Hours with the Stars* (1887), aimed not only at "popularizing scientific study" but did so by drawing attention to a proto-canonical collection of works by expert authors, such as the British astronomer Richard Proctor, who was fond of applying evolutionary ideas to heavenly phenomena, Spencer, Huxley, Tyndall, and Darwin.[39]

In both its technical and popular forms, evolution thus provided a vehicle for asserting the presence and role of a possible scientific orthodoxy. Before the twentieth century, however, the influence of such expert-driven science was not particularly robust and was relatively easy to resist. By the 1870s, evolution was drawing volleys of sometimes virulent criticism, much of it from outside the community of working naturalists. Controversy continued among men of science, though it frequently focused more on the mechanisms and limits of biological change than on its existence. The formidable and respected Agassiz had been replaced by a number of lesser-known critics, most of whom allowed for biological evolution in some form. Princeton's Arnold Guyot, for example, argued against a purely evolutionary account of the creation of matter, of life, and of humans, but he accepted natural development between these points of discontinuity.[40] Other critics included the American geologist J. W. Dawson and the British naturalist St. George Mivart. Mivart rejected natural selection, but not the possibility of a teleological, divinely directed evolution. However, his desire to combine his devout Catholic beliefs and his scientific work put him on the margins of the exclusive scientific community as defined by professional-minded observers such as Huxley.

Those who felt themselves pushed out of a threatened evolutionary orthodoxy also fought back by proposing their own versions of evolution that preserved the harmony of all truth and thus the communal right to make scientific judgments. A few interpretations of evolutionary ideas were given a supernatural edge by progressive theologians and by alternative religious movements, such as Theosophy or Spiritualism. Theosophy founder H. P. Blavatsky averred in her *Isis Unveiled* (1877) that a complete view of evolution began with "pure spirit, which descending lower and lower down, assumed at last a visible and comprehensible form, and became matter."[41] Meanwhile, the core of resistance to evolution as both a biological idea and as a broadly philosophical one increasingly came from outside the emerging boundaries of the scientific world. Darwin's ideas found little support in books by non-naturalists, from Princeton theologian Charles Hodge's *What Is Darwinism?* (1874) to the optimistically titled *The Refutation of Darwinism* (1880) by Warren O'Neill, a member of the Philadelphia bar. For other critics, the idea of evolution took its place within a much larger crusade against religion that included the spread of materialism, August Comte's positivism, new trends in theology that neglected the supernatural, and social forces and conventions that aimed to undermine traditional values.

Science and Not-Science

Debate over evolutionary ideas brought the tensions around science as a practice and body of knowledge into sharp focus. It also showcased Americans' struggle to talk about the emerging scientific establishment. During the early 1800s, most people depicted the world of knowledge as a more or less coherent whole. They expressed faith that the few disputed borders would quiet down as initial passions calmed and the number of recognized facts grew. By the last third of the century, the cultural maps that many people described were crisscrossed with fault lines between areas of science and not-science, a few of which had opened into widening gaps. Ultimately, these gaps jeopardized the unity of all truths that had formerly been an important measure of reliable and certain knowledge. Before the 1900s, however, most Americans were not ready to give up such unity. Bridges could still be built between science and other kinds of knowledge and a wide variety of American commentators on all sides of evolutionary debate labored mightily to build them. Simultaneously, while most observers worked to avoid total fragmentation, distinctions between kinds of knowledge could also be useful and so never totally disappeared. They allowed champions of science to display the advantages of their chosen field over other kinds of ideas. And they provided the means to limit the pretensions of scientific enthusiasts and to protect areas that were increasingly depicted as lying beyond the boundaries of science.

For instance, the scope of evolutionary claims varied widely, from the fairly moderate biological applications of Darwin to the truly cosmic ones of Spencer. In trying to determine the proper limits of evolution, if any, some Americans turned to an emerging distinction between two formerly synonymous terms: science and philosophy. Joseph LeConte routinely sought to protect evolution from the taint of unorthodoxy by maintaining that universal claims of materialism, and therefore any materialistic conclusions supposedly founded on Darwin's ideas, were examples of philosophy, not science.[42] Alternately, an 1878 article in the typically conservative *Catholic World* chastised evolutionists by noting that such inaccessible questions as the origin of human beings presented philosophical problems rather than scientific ones.[43] Spencer, who was widely seen as having transformed evolution from a matter of biology into a grand vision of the universe, frequently became a focal point of debate over differences between science and philosophy. Some of his admirers gladly accepted his conclusions as broad philosophical statements about the nature of the universe. E. L. Youmans claimed in an 1879 article praising Spencer that if done right, philosophy was as valid as science, and "in fact it is science in its highest form of expression."[44] Others sought to rescue Spencer from charges of philosophizing by enforcing a stricter division. The biologist David Starr Jordan claimed that Spencer's system was not just philosophy but "based wholly on the results of scientific investigation."[45]

At the same time that a separation between science and philosophy helped to manage debate over evolutionary ideas, it also opened larger questions about the relative value of these potentially alternative forms of knowledge. Jordan was particularly anxious to plant Spencer on purely scientific ground because, he asserted, while science provides useful and accurate knowledge, "philosophy is never identical with truth."[46] Youmans alternately found in a philosophical approach the promise of unifying increasingly diverse and specialized branches of scientific endeavor into a higher kind of knowledge.[47] At the same time, he was unwilling to give up an underlying connection between science and philosophy so that privileging one was not denigrating the other. A number of his contemporaries took a similar approach. Authors sometimes tried to balance separation with identity by calling philosophy the "science of sciences," in which the details of the individual "special sciences" converged.[48] There were also a few Americans who both expressed deeper divisions between the scientific and philosophical, and chose the latter as unambiguously more powerful and revealing. George S. Morris, recently appointed professor of logic, ethics, and the history of philosophy at the University of Michigan, noted in an 1882 issue of the *Princeton Review* that all final, exact knowledge about the world was really philosophical, not scientific. Science, as used in contemporary parlance, only involved the physical universe and drew a highly monochromatic picture of reality. Alternately, Morris wrote, philosophy's "organic-spiritual conception of the universe" could more easily understand society, art, and religion without reducing them to illusions.[49]

Evolutionists of various stripes and their critics sometimes turned to another term that was far less ambiguous in its value relative to science: "pseudoscience." Though talk about pseudo-science was still rare by twentieth-century standards, its appearance in magazine articles was increasingly common during the last quarter of the nineteenth century, in part driven by evolutionary debate.[50] Huxley wrote a number of articles in the mid-1880s, reprinted for American readers in the *Popular Science Monthly*, that put the distinction between science and pseudo-science on center stage. They were part of an ongoing debate with George Campbell, the 8th Duke of Argyll, who had risen to oppose Darwin's ideas in the 1860s. The duke argued for God's direct and careful control over his creation through His legislation and enforcement of natural laws. To Huxley, any inference of divine providence from the regularity of natural law was an invalid and pseudo-scientific one.[51] Huxley's use of the term aside, however, and perhaps contrary to the expectations of modern observers, most of the cries of pseudo-science did not come from evolutionists. Instead, they were made by those eager to protect real science from the influence of naturalistic evolutionary ideas. A columnist in the *Catholic World* depicted the Huxleys, Tyndalls, and Darwins of the world as the "modern Cyclops, who in forging their pseudo-sciences examine nature, but only with one eye."[52]

Though it was sometimes an attractive rhetorical resource for partisans debating evolution, pseudo-science varied considerably in meaning from one context to another. In his reaction to the Duke of Argyll, Huxley portrayed it as involving a supernatural interpretation of a purely natural concept. For critics of evolutionary ideas, by contrast, pseudo-science was often identified with an overly aggressive naturalism. In this latter sense, invoking the pseudo-scientific could actually help to smooth over potential conflicts between different realms of knowledge such as science and religion, and thus to a certain extent preserve a harmonious vision of undifferentiated truth. But even in that case, such harmony was secured at the cost of making scientific transgression especially noteworthy. Earlier in the nineteenth century, many Americans had understood that not everything that claimed to be science was and that some truths were false or incomplete when judged by what was considered to be scientific knowledge. But as long as science was a continuous part of and in many ways secondary to the larger world of truth, to be unscientific was simply to be untrue, and whether something was untrue mathematically, morally, economically, or scientifically made little difference. Pseudo-science was a rarely used resource in periodical literature. The growing popularity of the term over the 1800s distinguished scientific error, if on no other basis than giving it its own name, in turn suggesting something distinct and frequently special about the genuine science with which it was contrasted.

The high profile of evolutionary debate, like the potential scope of evolutionary ideas, highlighted yet another emerging boundary, namely, the one between scientific and popular knowledge. The influence of any rigid barrier between these two realms was still somewhat weak. Americans of many different kinds were perfectly comfortable speaking up about the merits or mistakes of evolutionary notions or proposing evolutionary accounts of phenomena themselves. Nevertheless, such a barrier was useful in ways that ensured its continued presence and even increased prominence. According to some would-be boundary makers, the language of the popular differed from that of the scientific in the purposes for which many Americans used it. For better or worse, popular expression most often sought to entertain. Science could be ennobling, uplifting, even interesting, but juxtapositions of the scientific and the entertaining—in the displays at the zoological gardens at Regent's Park in New York, or on the occasion of a new edition of Darwin's narrative of his voyage on the *Beagle*—were sufficiently remarkable to elicit special comment.[53] Rather than denigrate science, such a distinction could help to insulate it from the increasingly commercialized and occasionally sensation-seeking information industry that was growing up in the late 1800s.

One of the most common ways to separate the scientific and popular involved associating the one with accurate knowledge and the other with the absence of such knowledge at best and error and misperception at worst. An article on the critics of evolutionary ideas warned its readers that there was a "large class of minds even among those who esteem themselves as educated, who have no acquaintance with science and another, perhaps equally large, who have no ideas what is meant by the scientific spirit."[54] But antievolutionists were not the only targets of such sentiment. An author in the *Popular Science Monthly* claimed that the level of popular evolutionary ideas could be summed up in the idea that "most things 'growed.'"[55] The improper appropriation of evolutionary notions and their misapplication were only to be expected, another observer noted glumly, when people believed "every man his own evolutionist."[56] These claims cleared out a large space on the information landscape that was devoid of science and thus open to colonization and control by self-styled bearers of truly scientific knowledge. Former examples of carrying strictly scientific issues and debate into the popular realm became cautionary tales of boundary violation. Joseph LeConte reflected in 1887 that the *Vestiges* "was not a truly scientific work." Instead it short-circuited the process of proper science by making "an appeal from the too technical court of science to the supposed wider and more unprejudiced court of popular intelligence." Because the *Vestiges* aimed at reaching the general public, LeConte claimed, it was "far more eloquent than accurate; far more specious than profound. It was, indeed, full of false facts and inconsequential reasonings."[57]

Most distinctions between science and popular knowledge, however, did not posit any radical gap between them. Rather, they involved the creation of a kind of buffer zone that separated one from the other and that could offer safe passage between. This general idea was captured in the term "popular science," which was appearing with greater frequency in American magazines by the 1870s and after.[58] A columnist in a magazine called *Dial* attributed the work of tracing the impact of evolution on popular religion to people he called "buffers," who were not scientific practitioners themselves.[59] Popular science was not the only region where authors depicted science as amalgamated with what it was not. So-called "applied science" also began to receive attention during the late nineteenth century. Elevating applied science to the level of its own category suggested a new tendency to separate science as abstract knowledge from science at work in the world. Like its applied cousin, the concept of a popular science redefined an activity formerly considered an important part of the scientific enterprise—in this case the diffusion of knowledge—and made it supplementary. At the very least, crossing from one to the other implied a kind of simplification that tended to create a hierarchical relationship. A reviewer of one contribution to popular science singled it out as an example of "how scientific truth has been simplified for the people in our day."[60]

The notion of popular science, evolutionary and otherwise, became a vehicle for securing the rightful place of true science on Americans' cultural maps. Huxley received praise from an *Eclectic Magazine* columnist for combining the attributes of a man of science and of letters, who understood "that a scientific school can never become really powerful while it is content with the ear of the strictly scientific men." Scientific works especially written for children also had enormous value, "for science will never secure its due hold upon the thought and feelings of mankind" until it found a place in schools.[61] At the same time, popular science could create a gray area, in which scientific legitimacy was uncertain. Ironically, reaching out to a general audience in a lively and entertaining way in order to advance the cause of evolutionary science, and thus science generally, could appear to dilute or make light of real scientific knowledge. A reviewer of one of the *Half-Hour* books by Richard Proctor noted with concern that its tone was too flippant, which suggested "in it something besides a thorough and earnest devotion to his noble study. With such a man as Tyndall, the truth seeking to which he devotes his life seems too great a thing to permit him to chat about it as if it were the entertainment of the day."[62] A separation between the popular and scientific could even be turned against the most prominent evolutionists. A critic of Darwin compared him to "romance writers" who made it their business to create fantastic, though ultimately imaginary, scenes.[63]

Science and Religion

Of all the emerging examples of not-science that occupied Americans debating evolution during the late 1800s, religion was by far the most frequently discussed and the most controversial. Talk about the relationship between the scientific and the religious provided the clearest demonstration of the spread of well-bounded descriptions of science and attempts to preserve more diffuse and permeable depictions in the face of weakening consensus over the harmony of all knowledge. The violence of the imagery that partisans on all sides used suggested the keenness of their perceptions of change. Andrew Dickson White, formerly president of Cornell University, recalled that Darwin's *Origin* came into the "theological world like a plough into an ant-hill."[64] Mary Alden likewise resorted to destructive terms to describe the effects of scientific men on religious beliefs and institutions. Mention of "warfare" between science and religion was common. Many subsequent observers have used the notion of such a conflict to make sense of reaction to Darwin and evolution more generally. In fact, the real battle was not between science and religion so much as between different visions of science. The complexity of debate prevented any easy division between scientific supporters of Darwin, striving to purge their beloved science of every vestige of supernaturalism, and creationist opponents, who resorted to miracles and the immaterial world at every turn. Instead, debate over evolution involved struggle "between persons who wished to retain an older, theologically grounded science, and those who advocated a thoroughly positivistic science." Representatives of each side included naturalists and theologians.[65]

Although discussion of evolution was not a battle between two clear-cut camps, it did help to spread new ways of talking about science and religion as separate kinds of knowledge and distinct activities. The combination "science and religion," one of the most common and powerful means in modern parlance of implying such a separation, achieved widespread circulation in magazines during the 1870s and 1880s just as the evolutionary debate was heating up.[66] One basis of dividing the two involved linking science to the study of sensible nature and religion with supernatural or spiritual phenomena. Evolution provided a particularly potent example of this growing tendency. Earlier in the century, Americans seemed capable of assimilating natural physical explanations of the formation of the solar system and the earth or even, with the help of contemporary science talk, the outlines of human personality into traditional Christian doctrine. A nonmiraculous account of the history and perhaps even the creation of life, including human beings, offered a similarly difficult piece to fit into what had once been a harmonious system of truths about the world. But the means of talking about evolutionary science widely available later in the century only intensified the challenge. Charles Hodge complained that, under the influence of Darwin and others, science "was becoming more and more restricted to the knowledge of a particular class of facts, and of their relations, namely, the facts of nature or of the external world."[67] Appealing to phenomena beyond the natural could call down charges of pseudo-science, as it did from Huxley. The reestablishment of harmony was not impossible, but it did require nearly unprecedented amounts of rhetorical work that, ironically, helped to draw attention to a boundary between science and religion at the same time that it sought to bridge it.

The growing naturalization of scientific knowledge did not prevent some Americans from continuing to promote a less definite boundary between science and religion.[68] It remained possible even to deny any distinction between the two realms. An 1888 article in the New Englander by the Reverend Wilbur F. Crafts asserted in no uncertain terms that Christianity was indeed a science "since it consists, in its essentials, of proved knowledge, established, in part, like law and history, upon abundant and reliable testimony; established, in part, like the scientific certainties of gravitation and the roundness of the earth, upon a proved hypothesis" and "in part, like chemistry and medicine, upon repeated and thorough experiments."[69] Though few followed Crafts so far, the more modest claim that theology represented a legitimate branch of science appeared often enough to notice. So too did weakly bounded depictions of science that made definitions such as Crafts' possible. A correspondent in the morally uplifting and generally conservative Ladies' Repository disputed definitions that restricted "modern science" to the study of physical nature. Instead, science involved studying things with the use of reason and focusing on positive facts, whether in history, religion, philosophy, or nature.[70] Likewise, J. W. Dawson proposed in an article for Hodge's Princeton Review to use science in a sense that restricted its subjects to matter and force. But he stressed that he did not himself believe, with "one of our modern schools of thought," that such a restriction included "all science worthy of the name." Instead, he maintained that true science "is a term of wide application and may include any of those subjects of human thought in which facts are systematically arranged and referred to definite general principles."[71]

Yet broad definitions of science, past or present, were becoming more difficult to maintain in an environment increasingly rich with rhetorical affirmations of the separateness of science and religion. Some observers trying to make sense of the reaction to modern evolutionary ideas recounted the well-worn truism that new truths, whether those discovered by Darwin or promulgated by early Christianity, were always rejected.[72] Alongside such stories emerged an alternate version of events that attributed resistance to science to the conservative and often limiting forces of orthodox religion. John William Draper's *History of the Conflict between Religion and Science* (1874) and A. D. White's lectures and articles during the 1870s and 1880s, culminating in his *History of the Warfare of Science with Theology* (1896), told this tale to large audiences.[73] There were subtleties in both books. Draper tended to aim his gaze at the Catholic Church. And White professed to distinguish theology from religion. In either case, friction was often attributed to the errors and arrogance of religious authorities rather than to an inherent conflict between science and religion. Still, conflict of any kind helped to draw the eye and to make science a more singular and prominent entity. Perhaps the best example of historical conflict between science and religion became the now famous encounter between Galileo and the Catholic Church. Earlier Americans had certainly been aware of Galileo's troubles, but had tended to ascribe them to the same difficulties any new truth met. In the hands of many late-nineteenth-century authors, however, resistance to Galileo's work took on a particularly religious tone. Galileo himself became a martyr to the cause of separating science and religion, sometimes along with other notables such as Copernicus and Bruno.[74] In an overwhelmingly Protestant American culture, it did not hurt that the story had an anti-Catholic tone.

Like other emerging separations of science and not-science, distinctions between science and religion on whatever basis ultimately raised questions about the relationship between scientific knowledge and the more general realm of truth. One of the more common positions taken by magazine authors debating evolution involved asserting that science and religion revealed two distinct kinds of truth, a vision that historians have often called the separate-spheres model.[75] The frequent pairing of science and religion suggested as much by implying two binary categories that possessed a kind of ontological parity. In a discussion of evolution and materialism, Joseph LeConte related science and religion through the image of a shield, black on one side and white on the other.[76] The separate-spheres analogy became a dominant metaphor in popular rhetoric about science and religion, primarily because it helped to avoid conflict. William D. Le Sueur, a contributor to the *Popular Science Monthly*, wrote in 1885 that "all will be well if we keep everything in its own place, observing proper metes and bounds."[77] From this point of view, all signs of conflict between evolution and Christianity involved counterfeit science or error-filled religion, overzealous naturalists or stubborn theologians. Given the massive changes happening in American society and culture at large, no doubt it was comforting to find at least this much order in life.

But few Americans stopped there. Despite the fact that "divorce on the grounds of incompatibility has been constantly urged," most observers approached science and religion with the venerable notion of the unity of truth firmly in hand.[78] The alternative was for the "intelligent Christian" who believed in both evolution and the Bible to hold "the two truths side by side in his mind, carefully preventing any contact between them."[79] That seemed an inherently unstable condition. Instead, a host of harmonizers and reconcilers appeared amidst debate over evolution, including LeConte and John Fiske, eager to dispel the "phantom of the hostility between religion and science."[80] Sometimes such harmony required the help of a third party. An 1878 *Catholic World* article depicted philosophy as the "natural *umpire*" in the conflict between science and religion.[81] Most other harmonizers held firmly to the more direct claim that true science and true religion somehow synthesized on their own into one whole and consistent view of the universe. This depiction was implicit in LeConte's shield image. Numerous invocations of separate spheres were augmented by further affirmations of harmony and even alliance. A critique of Spencer's ideas declared that "all truth, whether natural or revealed, comes from the same Author, and must eventually harmonize, despite the efforts of science, so called, to play recklessly with faith, on the one hand; or over-zealous defenders of the faith to ignore scientific truth, on the other."[82] Observers on all sides of debate over evolution repeated the assertion that true science had done nothing but aid true religion or drawn attention to the religious value of scientific information. Mivart claimed in an issue of *Cosmopolitan* that study of the development of individual organisms showed that evolution must be guided by some higher plan, a view that culminated for many, even most Americans, in the notion of "theistic evolution" or evolution as God's method of creation.[83]

For others, the boundaries of science did not encompass a part of the truth, whether harmonious with other parts or not, but contained the whole of reliable knowledge about the world. From such a perspective, the continued commitment to the harmony of truths became a limitation, setting unacceptable conditions on the progress of scientific investigation. A rigorous assertion of separate spheres could help to prevent interference with the freedom of scientific research. Huxley caught the gist of this approach when he coined the word "agnosticism" to express his own supposedly neutral view, as a man of science, toward questions about God or the supernatural. He and others expected the same agnosticism in scientific matters from theologians. John Tyndall made some of the most aggressive comments about the freedom of science before the Belfast meeting of the British Association for the Advancement of Science in 1874. On that occasion he announced that modern men of science "shall wrest from theology the entire domain of cosmological theory" and that "schemes and systems which thus infringe upon the domain of science must, in so far as they do this, submit to its control, and relinquish all thought of controlling it."[84] Anything less, as one friendly reviewer of Darwin later noted, was "the spirit of the Inquisition."[85]

Behind Huxley's neutral-sounding agnosticism, however, was a further commitment to what modern historians often call scientific naturalism. Huxley, Tyndall, and a variety of other late-nineteenth-century men of science sought not only to free science from religion but also argued that traditional religion must be replaced or reconstructed on the basis of purely naturalistic science, which would become the primary and authoritative guide in human life. Huxley, for instance, preached his "Lay Sermons" and referred to the "Church Scientific."[86] Claims that science must replace religion were no doubt aided by the occasional assertion that traditional institutions and beliefs were in decline and sure to disappear soon. Its natural successor was a "'religion of humanity' or of science."[87] Even confirmations that science was supporting religion could soon slide into calls for the modernization or purification of religion through the application of science. Magazine authors friendly to new evolutionary ideas routinely made calls for a scientific religion "adapted to the day."[88] Scientific enthusiasts, such as the members of the Brooklyn Ethical Association, dedicated themselves to the application of science to every area of life, religion included. Widely noted books such as Balfour Stewart and Peter Tait's *Unseen Universe* (1875) and Henry Drummond's *Natural Law in the Spiritual World* (1884) seemed poised to extend the reach of science into intangible realms. While the science being exported in such cases was sometimes a purely physical one, it frequently included elements that would probably have made the John Tyndalls of the world cringe. Drummond's version of evolution, physical and spiritual, included the force of love as a major element, a scientific matter that Christianity had actually anticipated. Other versions of an evolutionary "spiritual science" were similarly eclectic and non-naturalistic.[89]

It was also possible to place the most significant kinds of truth about the world outside the boundaries of science, and thus to limit its supposed authority. This option did not require giving up a well-bounded image of science. Rather, such boundedness was accepted whole-heartedly. An author in an 1875 issue of the Presbyterian *Princeton Review* warned his readers that "we use the term, science, here with the limitation which such writers as Mr. Tyndall put upon it, meaning physical science." He then went on to show that such a science could never uncover the beginning of things, including life, and that the supernatural communication of facts about the history of life must be considered alongside purely natural data.[90] Thus the boundaries erected to protect science from outsiders could work as well to protect the rest of the world from science. The most common assertion about science and religion

among magazine authors discussing evolutionary ideas stressed the limited nature of science and its inability to "rise above its inferior" sphere. Science dealt
with how, not why, and when the "technically scientific task" of determining
basic physical matters was over, then science needed to "hand over her function to philosophy and religion and retire from the scene."[91] In the somewhat
more formalized system of John Bascom, previously president of the University
of Wisconsin, science presented the facts, philosophy interpreted them with
reason, and religion drew out their meaning for spiritual life.[92] Few expressed
the attitude so concisely as the classicist Tayler Lewis, who offered an early
declaration of this position in his *The Six Days of Creation* (1855). "We can get
along very well without geology," he wrote; "our intellectual and moral dignity
would not have been impaired had no such science ever existed."[93]

The Origin of the Scientific Method

Americans coming to grips with evolutionary ideas and their impact on other
areas of thought, including religion, frequently relied on methodology to describe what made science scientific. Talk about method was every bit as useful
earlier in the century too, but by the last quarter of the 1800s its character
had begun to change. Some science watchers, past and present, have depicted
the nineteenth century as witnessing the eclipse of a naïve and fact-obsessed
Baconianism by a synthetic and imagination-friendly view of how science
worked. Yet, the major development in late-nineteenth-century talk about
method occurred more or less independently of struggle over facts or hypotheses. Instead, it appeared in the emergence of ideas about a uniquely scientific
means of studying the world. Widespread methodological rhetoric had once
served to show what was shared by diverse areas of thought and the means by
which they participated in the unity of truth. That of the late nineteenth century, like a growing volume of contemporary science talk, increasingly became
oriented toward making science appear more distinct from what it was not.
Though rare before mid-century, references to something called "the strictly
scientific method, as it labels itself at present" began to spread in American
popular literature.[94] But, as usual, new boundaries also raised new tensions. It
was in the realm of method more than elsewhere that many Americans struggled to reconcile what would become the central tension in highly bounded
views of science: preserving a role for the scientific in the broader scheme of
society and culture while maintaining its separateness.

A simple inductive view of methodology was in fact under attack during the
last quarter of the century much more intensely than it had been before. Huxley
explicitly criticized a "Baconian" view of science as too limited. He claimed

that anyone familiar with the actual operation of science was "aware that those who refuse to go beyond fact, rarely get as far as fact." Likewise, the American philosopher and psychologist William James suggested that the Baconian believer might as well expect "a weather table to sum itself up into a prediction of probabilities, of its own accord, as to hope that the mere fact of mental confrontation with a certain series of facts will be sufficient to make *any* brain conceive of their law."[95] In some cases, evolution itself demonstrated the poverty of purely inductive method and the ascendancy of a newly powerful and singular scientific enterprise. The American essayist John Fiske claimed near the end of the century that Darwin's ideas prompted naturalists to abandon the "mood of the stamp collector" and rise to a higher kind of work.[96] Nor were advocates of biological evolution the only ones to express some discomfort with the perceived limitation of sticking to the facts alone. An article on the immutability of species in the *Catholic World* declared that "we have nothing in common with those who contend that the refutation of Darwinism lies solely with the mere compilers of fact-fanciers, florists, and breeders."[97]

Huxley and James did have a point, at least in light of contemporary talk about induction. Early-nineteenth-century public discussion of method was both more flexible and less obsessed with Bacon than later observers often recognized. Some of that flexibility remained later in the century too, helping to justify sometimes amorphous depictions of science. The notion that scientific knowledge was composed of facts organized by something as vague as common sense suggested that "any department of inquiry where a number of like facts can be collected and systematized may be made into a science."[98] However, widespread struggle over evolution later in the century tended to bring references to Bacon, appeals to facts, and claims of "severe" induction to the fore, perhaps even more than earlier in the century. Critics of Darwin, often writing in religious periodicals, relied on a fact-based, inductive view of science, presenting him and his supporters as "investigating nature, not in the interests of science properly so-called, but, consciously or unconsciously, to find facts to fit a hypothesis." If Darwin's ideas found general acceptance, one critic worried, "our own Baconian mode of viewing nature will be quite reversed."[99] Ironically enough, that reversal often involved a too obsessive concern for a limited class of facts. Other authors objected that any "mechanical, external, superficial, false" method threatened to "exalt the senses, which are the servitors of the mind, into the mind's masters." True science considered *all* the facts, including those of divine Revelation; true scientific method, which included the method of faith, aimed to read God's handiwork in these facts.[100] In such a view, all facts were "God's facts."[101] Similar depictions continued to forge strong links between science and other forms of knowledge, particularly religion.

Rather than be chained to facts, some aggressive defenders of Darwin turned toward rhetoric that cast the practice of science in a more creative light. In his address "On the Use of the Scientific Imagination," Tyndall argued that despite occasional excesses, society should be tolerant of the imaginative flights of geniuses such as Darwin rather than trying to limit them.[102] British economist Stanley Jevons' *Principles of Science* (1874) also posited a view of scientific methodology in which creatively generated hypothesis, even if the product of guesswork, played a large role. Such depictions echoed in popular rhetoric, too. During the 1880s especially, the term "working hypothesis" became common fare. New methodological luminaries also appeared in the scientific firmament. Beside such old stand-bys as Francis Bacon and Isaac Newton, both of whom most frequently symbolized proper induction and a focus on facts rather than theories, American proponents of evolutionary science sometimes also mentioned the late-sixteenth- and early-seventeenth-century astronomer Johannes Kepler. Antebellum Americans almost never mentioned Kepler, who first posited that planets orbited the suns in elliptical paths through a series of idiosyncratic leaps of imagination. Later authors did occasionally manage to tame Kepler enough to use him as yet another example of the value of induction, but in other cases, he served to demonstrate the value of more creative approaches to understanding nature, even through the use of sheer guessing.[103]

A creative view of scientific methodology could dramatically expand the scope and implications of scientific discovery. It could help to justify evolutionary accounts of life, which dealt with phenomena well beyond direct human observation. And it could offer a basis on which to found newly unified and distinct depictions of science. Henry Drummond asserted that for hundreds of years before 1859, science had "devoted itself to the cataloging of facts and the description of laws":

> Each worker in his own little place—the geologist in his quarry, the botanist in his garden, the biologist in his lab, the astronomer in his observatory, the historian in his library, the archeologist in his museum. Suddenly these workers looked up; they spoke to one another; they had each discovered a law; they had whispered its name. It was Evolution. Henceforth their work was one, science was one, the world was one, and mind, which discovered the oneness was one.[104]

But a less fact-centered vision could also blur scientific boundaries. If imagination were the key to truth, the geologist-cleric George Frederick Wright claimed, then the supernatural inspiration of Scripture taxed the imagination no more than some of Darwin's notions. The Reverend William Barry similarly concluded in an 1894 article that "the boast that science appeals to facts as its touchstone, while religion trusts to fancy, can no longer be maintained," because "both are seen to be the products of the intellect, equally valid or equally delusive." A synthetic and creative view of scientific methodology was particularly attractive to some figures beyond the pale of naturalistic science, such as Drummond. Theosophist Annie Besant, who lobbied for a more active method in modern scientific practice more in keeping with the mood of ancient Hindu ideas, sympathized with Tyndall's call for more imagination.[105]

Rather than follow Besant's lead or even Tyndall's, self-styled defenders of orthodox evolution most often returned to the familiar ground of facts to secure the boundaries of proper science. The editor of a magazine called the *Arena* wrote that true evolution does not "go beyond the facts, laws and processes or organic being." David Starr Jordan cautioned that "there is nothing 'occult' in the science of evolution. It is not the product of philosophic meditation or speculative philosophy. It is based on hard facts, and with hard facts it must deal."[106] More generally, defenders of Darwin's evolutionary ideas turned toward facts rather than potentially boundary-blurring invocations of the imagination. The Reverend George M. Searle agreed in the *Catholic World* with the British naturalist George Romanes that Darwinian evolution was "not by any means a mere theory," but rather "a simple statement of fact itself." Darwin was often presented, especially in science periodicals such as the *Popular Science Monthly,* as the cautious and unprejudiced searcher for facts, guided by "a system of the most severe inductive philosophy." Jordan noted that Darwin had only come to accept evolution when the theory became "an obvious inference from the facts." Even opponents of those "ultra-evolutionists" who had allegedly taken evolution too far in explaining the history of humanity or the origin of life sometimes acknowledged Darwin as the epitome of Bacon's ideal scientists or at least admitted that, though he was "over-fond of theorizing," science owed "him something for his collation of facts."[107] Though many critics of Herbert Spencer dismissed him as a deductive and overly speculative thinker, he too could find some shelter behind the shield of facts. A columnist in *Appleton's* noted that Spencer "had no favorite theories to support" in his "exclusive devotion to the pursuit of truth."[108]

Just as earlier in the century, most disputes over the methods of science focused on the nature of facts. But while the appeal to fact provided a route to defending scientific ideas and thinkers, in the more defined and segmented world of the late 1800s a focus on facts could also serve to enforce boundedness in ways that limited the scientific. A wide variety of authors used a physical-fact-centered methodology to confine a science that seemed as if it might be getting out of hand. The Reverend A. F. Hewit, a prominent Catholic apologist, sounded a common refrain when he reminded his readers that induction could not lead one to God. The astronomer could not see heaven. Yet, lamented another author, the devotee of science, "used to the precise terminology of science" in the material realm but unfamiliar with "sharp thinking or precise terminology in intellectual or spiritual phenomena," seemed to believe he could solve the mysteries of life and the universe from "within the narrow limits of his lab, among instruments of death." In such cases, people had a "right to demand that Science shall confine herself to facts." Another defender of orthodox religion declared, "Our scientists are seldom good logicians, and we have rarely found them able, when leaving traditional science, to draw even a logical induction from the facts before them." He continued, explaining that "this is wherefore they receive so little respect from philosophers and theologians, who are always ready to accept their facts, but, for the most part, unable to accept their inductions."[109] This view often implied a methodological gap between natural science, which relied on induction, and theology, which "as a science, is deductive."[110]

It was a short step from such comments to the belief that science involved a unique method, unlike that used in other areas of thought or activity. Earlier in the century, using a scientific method often meant simply being thorough and careful, without any particular connection to the study of nature. An author of a serialized work of fiction described a bank customer's impatience with the teller's "scientific method" of counting money as late as 1860.[111] Increasingly, however, caution and concern for detail came to be contrasted with being lively and engaging in a way that more firmly drew a line between the scientific and popular. An 1878 review of a work on birds complimented the author for "avoiding the technicalities of the ordinary scientific method" and thus exhibiting "the study of ornithology in its more agreeable and fascinating aspect." Numerous reviews of books on scientific subjects meant for wide audiences, all a part of the emerging genre of popular science, echoed such sentiments and praised the "art of combining a 'scientific method' with an animated exposition," so that such works were "neither too abstract nor too popular."[112] The potentially dull and strict requirements of scientific methodology sometimes heightened distinctions between science and art as well. An 1872 review of the work of actor and playwright James Mackaye evaluated his attempt to apply scientific method to a stage performance. Though the precise nature of his application was unclear, the reviewer did assert that the experiment was not entirely successful. Instead, he noted, "science had destroyed the artist. Rule and method underlie the efforts of every accomplished actor; but the moment they become apparent, the passion loses every touch and quality of genuineness."[113]

In sympathetic hands, the notion of a specifically scientific method helped to reflect the growing sense of science as a unified endeavor rather than a collection of branches. Science "one and indivisible" demanded a single method.[114] For some Americans, this purportedly unique method provided a new source of scientific power and a new means of exporting the influence of science across its boundaries. In 1867, Youmans criticized what he felt was a widespread notion that the "scientific method of inquiry"—by which he seemingly meant induction—could not be applied to the study of humans. Needless to say, Youmans felt such application to be crucial. Eleven years later, a reviewer in the *Overland Monthly* praised the further extension of "the scientific method to literature."[115] But a specifically scientific method could also restrict the reach of science. In fact, most explicit invocations of scientific method during the late 1800s occurred in this light. In his *Natural and the Supernatural* (1858), Horace Bushnell warned readers of too much confidence in the methods of science, "as if nothing could be true, save as it is proved by the scientific method." Instead, "the method of all the higher truths of religion is different, being the method of faith; a verification by the heart, and not by the notions of the head." Similarly, an author speculating on the future of human character in the *Ladies' Repository* reflected that "every generation, as it accumulated fresh illustrations of the scientific method, is more and more embarrassed at how to piece them in with that far grander and nobler personal discipline of the soul which hears in every circumstance of life some new word of command from the living God."[116]

A New Scientific Species

The methodological fractures between kinds of knowledge that debate over evolution highlighted sometimes spread to depictions of those who pursued science. Claims about how science worked were in fact intimately related to the ways Americans talked about who was doing it—that is, to paraphrase William James, whether scientific practice was open to just *any* brain or whether there were limits some people could not cross. Restricting science to a particular class of facts, for instance, could privilege certain sorts of scientific hopefuls over others. Any literate American could gather facts about creation from Scripture. Not everyone had access to the most recent fossil discoveries. The spread of more creative, hypothetical portrayals of method also made the ability to observe a few facts of whatever kind insufficient qualification for claim to scientific standing. Such portrayals had implications for the construction of hierarchies among scientific practitioners as well. Now and again, some proponents of evolution talked about the division of labor between the observer, "who contents himself with merely ascertaining facts," and the thinker, "who gives shape to science."[117] Closing off widespread access to scientific participation in practice and in principle was a crucial prerequisite to the formation of a powerful scientific establishment, as was the diffusion of the rhetorical tools for talking about such a thing. At the same time, as in the case of method, erecting boundaries to keep outsiders away from science were equally useful for containing scientific practitioners and limiting the scope of their claims.

A growing perception of those who did science as a distinct and separate group appeared most clearly in the diffusion of several important rhetorical tools. Partisans on all sides of evolutionary debate sometimes took advantage of increasingly common appeal to the "scientific world"—on one occasion even "the scientific world and his wife."[118] E. L. Youmans sought to distance the work of Spencer from the similarly cosmic evolutionary scheme that had appeared in the *Vestiges* by noting that the latter was so crude "as not to be accepted in the scientific world."[119] The same tactic could be turned against evolution itself. An 1863 article in the *Princeton Review* claimed that a proposition about nature acquired authority when the "scientific world accepts it without question," a criterion that evolutionary ideas, Darwinian and otherwise, had not met.[120] The tendency to restrict scientific participation reached a kind of apex in use of the term "scientist." Unlike popular alternatives such as the relatively amorphous "man of science" or discipline-specific labels, "scientist" provided a term that seemed to stress the unity of science rather than its fragmentation and to distinguish practitioners by their focus on science rather than their cultivation or character. An 1875 reference to "the so-called modern 'scientists'" indicated how unfamiliar the term sounded to some people.[121] But its use in American magazines rapidly increased after 1850, eventually eclipsing both field-specific designations, such as "astronomer" or "geologist," and "man of science."[122]

Debate over evolution itself contributed to discussion of a contemporary scientific community by drawing attention to active scientific men and their opinions. A tongue-in-cheek critique of travel by railroad noted the large number of vendors selling literature on trains and the many people who bought their wares. "Pile under your chin," the author wrote, "a goodly batch of Huxley, Tyndall, Lubbock, Mivart, Proctor, . . . and other popular writers on science, and you will find a customer."[123] Still, older ways of talking survived. Many descriptions of scientific practitioners during the late 1800s did little to distinguish them from any other seeker of truth. Mention of institutional affiliation to establish scientific standing was almost completely absent from articles on evolution in American magazines. Instead, some authors emphasized the amount of knowledge about and degree of study of a topic. Few practitioners seemed to demonstrate the traditional and generally accessible characteristics of the true man of science better than Charles Darwin.[124] Just as some supporters made Darwin an archetype of inductive thinking, others frequently lauded him as a good person and an exemplar of the high moral fiber that doing science demanded. Particularly after the posthumous publication of his *Life and Letters* (1887), Darwin appeared as a thoroughly ordinary person with a "grandly simple rectitude of character."[125] The model into which such admirers put Darwin appealed more broadly as well. An 1871 notice in *Appleton's* reported on the trial of Edward Rolloff, who had been accused of murder. In his own defense, Rolloff pointed to his interest in "science and the higher departments of literature," assuming that it demonstrated that he was "actuated by sound moral principles and incapable of great crimes."[126] Rolloff's defense failed, but subsequent authors sometimes echoed his claim that all genuine scientific investigators showed such traits as scrupulousness and honesty.

The identification of science with knowledge organized by common sense also continued to depict science as potentially open to everyone. The link between science and commonly held habits of thought was strong enough during the late 1800s that even Huxley, despite his criticism of naïve Baconianism elsewhere, used it. He noted in *Darwiniana* (1897) that "the method of scientific investigation is nothing but the expression of the necessary mode of working of the human mind." He claimed that the difference between the "mental operation" of the ordinary person and the man of science was no greater than "between the operations and methods of a baker or a butcher weighing out his goods in common scales and the operations of a chemist in performing a difficult and complex analysis by means of his balance and finely-graduated wits."[127] Huxley did not make the further claim that science was or should be open to everyone, but others did. In his *Introduction to the Elements of Science* (1894), Mivart stressed that "science is nothing more than plain reason and common sense used in a methodical manner and applied to the examination of various objects around us." He informed his readers that anyone who possessed knowledge and desired more "already possesses the scientific spirit, and nothing but a little patience and perseverance are needed for him, sooner or later, to become a true man of science."[128] Likewise, a columnist in the *Ladies' Repository* claimed in 1874 that "every field of human thought ought to be traversed and re-traveled by all kinds of minds and degrees of talent, until the true is saved and the false is thrown away."[129]

Alongside such broad descriptions, however, there was an increasing tendency among Americans promoting or contesting evolution to distinguish between devotees of science and other groups. Just as science and religion often became binary categories, so too did naturalists and theologians. Not everyone enforced such a distinction. The Reverend Thomas Hill noted in *Bibliotheca Sacra* in 1878 that truths, including those about nature, were more likely to be found by those looking "with reverent awe."[130] By 1891, an article in the *New Englander Magazine* asserted that while people needed to be morally upright to perceive religious truth, they did not need to be in order to understand science.[131] Proponents of a strong and autonomous scientific community were often particularly keen to separate men of science and theologians who, as Huxley claimed, were increasingly depicted as "citizens of two states, in which mutually unintelligible languages were spoken."[132] Likewise, though some observers continued to maintain that "our greatest scientists are also our greatest philosophers and the reverse," others began to tease the two categories apart. Asa Gray took special care to note that though philosophers of evolution had been common, Darwin was not among them. "He was a scientific investigator," explained Gray, "a philosopher if you please, but one of the type of Galileo." Another author, reflecting on Huxley's labeling of Darwin as a philosopher, explained that "we are to understand the term as then used to mean what is now generally called a man of science rather than a metaphysician."[133] Once again, Herbert Spencer often occupied center stage in wrangling over the difference between philosophers and men of science. And though his position was far from clear—his friends and enemies classified him as both, one or the other, and neither at different times—the very desire to determine his true nature suggests the growing distinctness of the two identifications.

The creation of popular science also allowed Americans to make fine divisions among types of people. The language of "professionals" and "amateurs" began to appear amidst struggle over evolution, sometimes with the latter in a negative light.[134] Meanwhile, some men of science disputed the right of "popular" writers to be considered scientific insiders.[135] Advocates of self-consciously orthodox brands of evolution frequently complained about the superstitions and misconceptions promulgated by "magazine philosophers" and "newspaper scientists."[136] Such charges could even target some of the leading lights of contemporary science. In 1878, Dawson contrasted Tyndall's careful experiments and original research with the "quaint and ruder imaginings of the same mind when, in the presence of popular audiences, it speculates on evolution." The Reverend George McDermot took the much stronger view that, though Tyndall was popularly considered a great scientific man, he was really just a lecturer. Traditional science, another observer noted, worked calmly and without fanfare. But, in the modern day, he complained, "Huxley cannot anatomize the leg of a spider without publishing the process in the newspapers, with some reflection upon its bearing and probably fatal effect upon the Mosaic record." Such desire for sensation might have been forgivable in fiction but was inexcusable in science.[137]

Not everyone was happy with the growing boundaries between those who practiced science and those who did not. One columnist, reacting to a description of the ideal scientific practitioner that was added to the printed version of Tyndall's Belfast address, objected to any depiction of the "true man of science" as if "he were a being of superior mold." Instead, the author suggested, when the desire for truth "is claimed as the peculiarity of any one class of men on the face of the earth, we smile, and put the claim to the account of the vanity of its authors."[138] A less accessible, less public scientific community also raised the possibility that science might be hijacked by a small group of agitators. And indeed, champions of orthodox Christianity did sometimes express concern that an "anti-Christian sect" or cabal, with Huxley, Tyndall, Spencer, Darwin, and friends at its core, was perverting science for its own atheistic ends.[139] Elsewhere, critics of evolution often charged that its advocates, and members of the scientific world more generally, had come to behave as a "priesthood," demanding a faith no less blind than the one they criticized among religious believers. An author in the *Bibliotheca Sacra* complained in 1872 that "men of science have come to occupy something of the position once held by the religious teacher, and to feel that whatever they taught would be accepted by the hearer as certain truth."[140] Several years later, a columnist in the *Unitarian Review* railed against the arrogance of scientific men who "undertake to excommunicate from the scientific body every man who is unable to close his eyes against the clear manifestation of the presence of an intellectual plan in the organic world" and "to overbear our judgment on the theories they propound, and the religious doctrines they advance, by telling us that modern science says so and so." Such behavior was as bad as that of the authoritarian Catholic Church.[141]

As with methodology, a strong distinction between the scientific and nonscientific worlds could be as useful at keeping scientists confined as at keeping outsiders out. Narrowness ranked first among the attributes of scientific practitioners in magazine articles discussing evolution.[142] Words such as "cold," "dry," or "mechanical" frequently appeared in descriptions. An 1890 column in the *Overland Monthly* noted that, while classicists did not often see the value of science, the other side represented "a class of men whose heart never beat in delight over a charming poet, men who cannot appreciate a literary master piece because it cannot be measured, calculated, proven, because it is not an exact discipline."[143] The biographies of first John Stuart Mill and later Charles Darwin were frequently used to illustrate this point. Reviewers routinely described their impressions of Mill as unemotional and "all intellect." In Darwin's case, a casual comment that he had lost interest in art and

poetry as he aged became one of the most cited and discussed passages of his
Life and Letters. An 1888 *Atlantic Monthly* article ascribed this atrophy of his
aesthetic tastes to a rejection of all other routes to knowledge except through
the study of the natural world. The author noted that Darwin himself was "so
convinced that his life had been narrowed in these ways that he says if he had
it to live over again he would have planned to give a certain time habitually to
poetry."[144] Sometimes these same images made their way into light literature.
An 1886 short story that featured a state entomologist described him, with
a phrenological flourish, as having a head whose back was flat "as becomes a
man given to the severe application which must not be distracted by patriotic
or domestic side-issues." While he becomes involved briefly in a love affair, he
is soon jilted and consoles himself by working for a bill in the state legislature
to grant half-a-million dollars to entomological research.[145]

Depictions of scientific practitioners as limited and narrow condensed
particularly around the label of scientist. Such narrowness could be useful.
An 1871 article in *Ladies' Repository* defended Darwin from the charge of
being anti-Christian by noting that "he is simply a scientist, unconcerned
about God or religion."[146] Likewise, in a published debate between the astronomer Simon Newcomb and several prominent theologians, Newcomb
claimed that he was not qualified to judge the impact of evolution on religion. Instead, "scientists" studied the relations of natural phenomena. All
else he left "for investigation by other methods than those with which he is
conversant."[147] In truth, Newcomb possessed a strong bias against orthodox
Christianity. His appeal to the impartiality of the scientist helped to conceal
this.[148] But narrowness could also be used to confine the potential authority of newly christened scientists. The theologian Noah Porter suggested in
1880 that "the scientist" was not well versed in dealing with phenomena
outside of physical nature and so was often led astray in matters of the mind
or spirit. An author in the *Catholic World* made the point even more strongly,
wondering that while "it is the fashion to write bold words on the liberty of
the scientist," who could imagine "limiting him further?"[149] In one of the
more forceful assertions of this attitude, a columnist in 1873 claimed that if
someone argues that "scientists accept the theory" of evolution, "and that we
therefore should, we reply *mere* scientists do; and of all men, the least safe of
guides is the mere scientist."[150]

Survival of the Fittest

Over the last half of the nineteenth century, the basic tools of American science talk gave authors increasingly effective ways of making science distinct.

Habits of speaking that distinguished scientific knowledge and practice from other realms, whether philosophical, popular, or religious, sharpened science's shadow. Claims that it operated by a special method all its own or that it was represented by a particular and exclusive group of practitioners inscribed its boundaries more prominently on cultural maps. Such rhetoric would become the norm during the next century. During the late 1800s, however, it was still controversial. Like the picture of nature that some Americans drew from evolution, contemporary talk about science was often "red in tooth and claw."[151] On a wide variety of fronts, newer ways of talking about science clashed with more traditional ones that deemphasized strict scientific borders in favor of more amorphous divisions and stronger connections, even identities, with other kinds of truth. Nor was such struggle only about the nature of science; it also involved the role of science on the larger landscape of American culture.

With distinctness came a sense of privilege over other kinds of knowledge. Numerous observers, for instance, worried about the authority of a newly coherent and less accessible scientific community. We have already seen such worry in charges that scientists formed a new kind of priesthood. Others, often those aspiring to be intellectual elites themselves, complained that the unlearned members of society considered the authority of modern men of science as sufficient evidence for any proposition. And an 1876 correspondent in the *Eclectic Magazine* went out of his way to emphasize that the statements of scientific men were not scientific facts, despite the tendency of even educated people to accept them without reasoning or reflecting for themselves. Even Alfred Russell Wallace worried, unnecessarily as it turned out, that the "masses" will "blindly adopt" natural selection without giving it the thought it deserved.[152] The presumption, sometimes even fear of the authority of scientists did not go without challenge. In 1893, the Reverend H.I.D. Ryden advised "a little more confidence in science—a little less confidence in scientific men." But even the reverend admitted that "of science, of accurate knowledge, we cannot have too much."[153]

Other observers also extended enormous authority to science. "The gigantic strides of modern science," Professor Leon C. Field informed the readers of the *Ladies' Repository* in 1873, "the astonishing revelations of modern philosophy, the wonderful discoveries of modern enterprise, the marvelous and manifold inventions of the present age, are themes of unlimited panegyric."[154] By the 1880s and 1890s, it was not uncommon for authors to reflect that the "nineteenth century has been an age of science" or that they lived in a "scientific age" in which the "scientific spirit" was gaining in strength. Henry Drummond claimed in his widely noticed *Natural Law in the Spiritual World* that the "scientific demand of the age" required that everything concerning life and conduct be placed on a scientific footing. Similar declarations were common enough to be parodied. A reviewer of military novels in the monthly *Catholic World* quipped that because "our day is pre-eminently 'scientific,'" the time was fast approaching when a person "will have to pull off his boots on strictly scientific principles, and with the least expenditure of force, lest that act may disturb the exquisite poise of the cosmos." With value came the occasional fraud. A correspondent in the *New Englander* sounded a note of caution in 1888 by pointing out that "nothing is more frequently counterfeited than the word 'science.'"[155]

To take such impressions at face value, however, would be to ignore the larger picture of which they are a part. The idealization of science increased most noticeably in certain areas of the intellectual world, particularly the emerging research university.[156] But any claims about a universal deference to science in all things reflected the spread of ideas about a passive and almost childlike public rather than the reality of its acceptance of scientific ideas. Concerns about the authority of scientists were sometimes linked to a general worry about the influence of the mass media. An 1873 article in *Ladies' Repository* noted in particular the tendency of many magazine readers to "swallow any bolus that speculative doctors of chances may drop into their gullets," especially when an article begins with "Dr. Dumkopf says."[157] Joseph LeConte complained in 1885 about notions "attested to by newspaper scientist, and therefore not doubted by newspaper readers."[158] Still, most Americans did not become agnostics simply because Huxley was one. Science could provide a source of rhetorical power, particularly as its more distinct appearance drew the eye, but its boundaries were still in considerable dispute. Given its uncertain position in an American society being transformed by powerful economic, political, and social forces, declarations about the massive authority of modern science may have been less about the fact that science was significant than about the anxiety that it should be.

That anxiety sometimes grew in light of the very separateness that was making it possible to invest science, by itself, with increased prominence and independence. Concern over a growing gap between science and the general culture of which it was once an important component appeared among many observers, who drew worried attention to what they saw as the tight focus of younger naturalists. In an 1898 article in *Science* celebrating a half century of evolution, biologist Alpheus Packard expressed his concern that the "experts in the use of the microtome and of reagents appear to have but little more general scientific or literary culture than high-class mechanics."[159] A. D. White recalled in 1878 the antebellum days when men of science happily traveled the lecture circuit. Sadly, he wrote, the present generation of scientific workers

often underrated "everything except minute experiments."[160] Likewise, John
Fiske noted that to the specialist "'science' means the collecting of polyps, the
dissecting of mollusks, the vivisection of frogs, the registration of innumerable
facts of detail without regard to the connected story which all these facts,
when put together, have it in their power to tell."[161] Many scientists did in fact
display an increasingly inward focus on their work as the nineteenth century
ended and the new century began. While the great popularizers such as Huxley
and Tyndall sought to intensify the prestige of science through appeals to the
general public, professional scientists of a later generation seemed to succeed
in making a place for their research by abandoning an abiding concern with
widespread outreach and focusing on those with power and money, such as
wealthy industrialists. An 1876 "Editor's Table" in *Appleton's* presciently urged
elderly philanthropists to leave their money to support men of science and
free them from the burden of both pursuing their science through research and
having to reach out to popular audiences.[162]

As we have seen again and again, while it became increasingly possible for scientific researchers to turn away from the public at large, it also became increasingly possible for those content to remain outside the scientific
community to confine the impact and place of science. Though Susan Faye
Cannon's assertion that natural science formed the norm of truth during the
early Victorian period may not be in evidence in American popular rhetoric,
her sense that science lost something over the course of the late nineteenth
century is reflected there. Science could not really become a privileged kind
of knowledge until it was clearly separate from other realms. It could then
offer to unify the expanding and fragmenting American cultural landscape
under a new, specifically scientific banner. And indeed, some Americans made
the choice to use science in this way. But it could also become just another
kind of information among many others, and often not the most important
kind. Psychologist G. Stanley Hall claimed in 1882 that children should be
taught other truths than science. To do otherwise would risk making the mind
"pragmatic, dry, and insensitive or unresponsive to that other kind of truth
the value of which is not measured by its certainty so much as by its effect on
us."[163] The choice to invest science with authority and significance became
just that—a choice. And public rhetoric provided ample tools for defending
one's inclinations either way.

Cannon suggested that the debates over evolution helped to create a
"multi-normative world," in which many disconnected truths could exist
side-by-side. Still, popular debate over evolutionary ideas was hardly alone in
prompting such a change. In an 1869 article in *Appleton's* called "The Burden of Knowledge," Youmans suggested that Darwin's struggle for life and the

"battle of races" had a counterpart in the world of knowledge.[164] Facts and ideas seemed more widely available than ever before, a situation noted by other authors. Earlier, such cultural activities as natural theology had helped to invest science with a transcendent importance and to make it worthy of consideration. But a bounded and independent science was deprived of such links, and with them any necessary imperative, moral or otherwise, to demand the attention of readers with many more things to think about. At the same time, the spread of information was increasingly shaped by a commercialized mass media, which tended to shift an individual's relationship with knowledge toward more consumer-oriented models. For some Americans, science remained a kind of religion, resonating with power and prestige. For many others, its separateness made it limited and unappealing. In either case, science increasingly became another consumable in an expanding marketplace of ideas.

As the twentieth century dawned, evolution itself met somewhat different fates, depending on which marketplace it was being peddled in. Among those increasingly laying claim to the label of scientist, there remained considerable controversy about the precise mechanisms through which organisms evolved. Darwin's natural selection continued to be an unpopular choice. Most naturalists preferred Lamarckian explanations that to varying extents stressed organisms' active adaptation to their environments. During the early 1900s, however, natural selection was amalgamated with the rediscovered work of Gregor Mendel, an Austrian monk who proposed a theory of heredity based on genetically determined traits. The so-called Evolutionary Synthesis of the 1930s and 1940s, which brought together research in a variety of fields, ultimately made Darwinian natural selection synonymous with evolution.[165] Yet, at no point was biological evolution itself questioned by even a significant minority of working biologists. Outside of established scientific circles, meanwhile, evolution of whatever kind frequently occupied what seemed to be a state of permanent controversy.

Though debate in the years immediately after 1900 was subdued, it soared to new heights during the mid-1920s when some fundamentalist Christians sought to ban the teaching of evolution in public schools. Their crusade reached its most newsworthy form in the famous Scopes trial of 1926. The trial, engineered as the test of a Tennessee state antievolution law, pitted one of America's foremost orators—William Jennings Bryan—against one of its most successful litigators—Clarence Darrow—becoming one of the first modern media circuses in the process.[166] The antievolution crusade of the 1920s sent partisans on all sides reaching for well-worn tools of science talk. The rhetoric it unleashed largely focused on the same issues as late-nineteenth-century evolutionary debate, especially distinctions between science and religion and between scientific and popular understanding. Discussion reflected the changing nature of that rhetoric, but it did not press many Americans to really grapple with new ideas about the nature of science. Rather, it was up to another scientific development to prompt many Americans to plumb the depths of their rhetorical reservoirs, namely, Albert Einstein's theory of relativity.

Chapter 3   Relativity

A Science Set Apart

In EARLY 1930, NEWS of a riot in New York City circulated widely. That was, perhaps in itself, not particularly noteworthy. But this riot occurred at the American Museum of Natural History. And, more unusual still, it began when several thousand "men, women and children" tried to get into a small theater where a film explaining Albert Einstein's theory of relativity was going to be shown.[1] In fact, the "mob" was actually a gathering organized by the Amateur Astronomers' Association of New York. Police had been called, said the Association's treasurer in a letter to the *New York Times*, not to control a riot but to help as ushers with the larger-than-expected audience. This letter apparently did little to set matters straight. The claim that the Einstein film had begun a riot possessed a deep-seated appeal. It continued to circulate, for instance, in an article in *Scientific American* later that same year and in a 1932 popularization of relativity by H. Horton Sheldon, professor of physics at New York University.[2]

One reason rioting over relativity was so easy to believe was that few contemporary scientific developments drew as much attention. The well-known engineer Charles Steinmetz declared that relativity was "the greatest scientific achievement of our age."[3] Others celebrated it as an unprecedented breakthrough in human understanding. Despite its technical complexity, relativity also became the subject of headlines. As early as 1923, the British biologist Julian Huxley claimed that relativity was "in the air. It is so much in the air that it becomes almost stifling at times."[4] A few years earlier, the journalist Edwin E. Slossen recalled an encounter between himself and another passenger on a streetcar who was engrossed in a newspaper article on relativistic physics. The conversation that ensued prompted him to reflect, with a little exaggeration, that "our newspapers are sending out their reporters to interview astronomers as well as actresses and devoting pages to speculations on the nature of space and time as well as on the state of the market."[5] In subsequent years, newspaper and magazine reporters scrambled to cover the latest tests of relativity, while Einstein himself became a media celebrity.

Discussion of relativity, like phrenology and evolution before it, also provided the opportunity for Americans to talk about science. Supporters widely depicted Einstein's work not only as the epitome of the latest and most important research, but also as a revolution in the nature and operation of science itself. Critics claimed that the new physics was turning science into mathematics or, worse, philosophy. Observers on all sides used the occasion of widespread attention to relativity to comment on the relations between science and the general public. Yet there was somewhat less explicit discussion of the nature of science during the 1920s and 1930s than there had been during the last third of the nineteenth century. In part, this was because the jumbled and tumultuous notions about the nature of the scientific that jostled together during the 1870s and 1880s had given way to greater stability and more widespread conviction that, whatever science was, it was something distinct and recognizable. By 1940, Einstein could confidently—if overoptimistically—assert that "it would not be difficult to come to some agreement as to what we understand by science."[6] Science had basically come to assume its present-day form, focused on a constellation of disciplines devoted to the naturalistic study of the physical world, and no longer contained elements that would be unfamiliar to the modern observer. There were still disagreements about what science was and was not, how it worked, and who could speak for it, but increasingly they occurred on the margins where boundaries seemed most in danger of being transgressed.

In keeping with such changes, American science talk had become stable, not so much in detailed content, but around some of the important science-defining keywords that were emerging at the end of the 1800s. Images of science in popular discussion increasingly revolved around reference to its unique and potent "scientific method," the distinctions between "science and religion," or the association of science with a special group of authoritative "scientists." Older, broader visions of science continued to exist, but they became less and less visible in popular discussion. The situation was in some ways the reverse of the early nineteenth century, when the general consensus about amorphous definitions of science was shadowed by suggestions of less permeable scientific boundaries. And, as usual, ways of talking about science that depicted it as more distinct invested the scientific with both a new sense of power and also one of remoteness. They allowed authors to assert the advantages of science over other forms of knowledge. They also suggested the increasing inaccessibility of scientific truth and even its limits in comparison with other aspects of American culture.

Contemporary depictions of science had "riotous" aspects as well. Some Americans, chafing against restricted depictions of science, focused more and more on the possibility of reestablishing connections between science and the larger world of knowledge. The popularization of relativity provided the most ambitious early-twentieth-century attempt to show how seemingly abstract and remote scientific knowledge was deeply connected to ideas beyond the borders of science proper. William Wunsch observed in a 1931 issue of *New-Church Review* that the "general idea of relativity seems to have leaped out of the field of time and space where Mr. Einstein has displayed it into the field of the timeless, the moral and religious."[7] And yet, as the rhetorical boundaries of science grew in strength, they also became easier to transgress. Ronald Tobey argued that the supposed incomprehensibility of relativity helped to destroy the vision of science as broadly relevant to American culture.[8] But just as evolution did not single-handedly destroy the early-nineteenth-century "truth-complex," the increased sense of remoteness of scientific ideas emerged from trends toward more bounded depictions of science already visible during the late 1800s. In 1921, the British philosopher and politician Richard B. S. Haldane warned that "science is apt to find itself in strange regions if it does not limits its scope with genuine self denial."[9] As the separateness of science grew, technology became the safest and most reliable means with which to link the scientific and the world of ordinary experience.

The Famous Dr. Einstein

In 1926, the British physicist Oliver Lodge mused that "half a century ago, or perhaps less, evolution was the word to conjure with. Now it appears to be relativity." Rhetorically, at least, early-twentieth-century relativistic physics seemed to be following in the footsteps of late-nineteenth-century evolutionary biology. In 1921, Frederick E. Brasch, a Chicago librarian seeking to compile a bibliography of works on relativity, announced that "not since the doctrine of evolution was promulgated, has any advance of intellectual progress, either of philosophic or scientific importance, caused such a profound interest, popular or scientific." As time went on, Einstein himself became the focus of much of that interest. A reviewer of an early Einstein biography noted that "in his humility, his shyness, his lack of desire for controversy, his generosity and simplicity, he reminds one strikingly of other great scientists, particular Charles Darwin." Opponents of relativity sometimes used Darwin as one yardstick for measuring true science, as well. Charles Lane Poor, an astronomer at Columbia University and a determined critic of relativistic physics, suggested that the rapidity with which advocates declared relativity proven contrasted with the caution of the "accepted scientific method" as demonstrated by Darwin. The English naturalist worked on evolution for twenty years before publishing his ideas. Nor was Darwin the only historical figure to be mustered. Without meaning it as a compliment, Lodge characterized the philosopher R.B.S. Haldane as Einstein's Herbert Spencer.[10]

Despite such comparisons, the scientific landscape on which discussion of relativity took place looked very different from the one evolution had encountered. Just as in a wide variety of occupational areas during the early 1900s, from medicine and engineering to education, the once nascent scientific establishment matured into an increasingly organized system of practices and institutions that firmly distinguished between insiders and outsiders.[11] Graduate training, specialized journals, jobs in academic and industrial laboratories, and membership in exclusive professional organizations all inscribed deeper divisions between members of the scientific community and those whom the leadership deemed amateurs.[12] Such developments allowed researchers to extend scientific knowledge to an unprecedented degree. But they also restricted many Americans' ability to do science themselves or at least to contribute anything of value. The amateur tradition continued in a number of fields with a large observational component, most notably astronomy. Amateurs sometimes formed their own institutions, such as the Amateur Astronomers' Association of New York. But generally, their work was either marginalized by the established scientific community, which typically had a very different set of interests and goals, or was mobilized to be of service to professionals, a trend already underway in such popular endeavors as botany during the late 1800s.[13]

Relativity took its first steps toward entering this new scientific landscape in 1905, when Einstein published his first relativistic paper in the German physics journal *Annalen der Physik*. There, he provided a startling answer, both in its conceptual simplicity and radical implications, to problems regarding motion that had developed in late-nineteenth-century physical science. The famous Michelson-Morley experiment, though it now appears to have had little influence on Einstein himself, embodied at least one aspect of the difficulty. The best-known version of the experiment took place in 1887 in Cleveland, Ohio, where Albert Michelson and Edward Morley had tried to detect the motion of the earth through the ether, the ghostly ocean that most scientists believed filled all space and through which light waves supposedly traveled.

Contrary to all expectation and common sense, their equipment registered nothing. In subsequent years, several researchers offered possible explanations, and certain details of those explanations anticipated Einstein's eventual solution. But where others continued to work within the existing framework of ether-based electromagnetic theory, Einstein bypassed the existence of the ether entirely and simply affirmed the relativity of all motion, including that of light. Soon unhappy with the restricted nature of his initial ideas, Einstein sought to extend relativity theory to cases of acceleration, including the action of gravity. After a decade of hard work, he published a preliminary form of his general theory in 1916, and then offered a more complete and slightly corrected version in 1918.[14]

Einstein's generalized equations ultimately provided the key to bringing relativity to broad public notice. Those equations predicted that light passing through a gravitational field would be deflected. In any terrestrial laboratory, however, the degree of that deflection was extremely slight since the earth's gravity was much too weak to have an observable effect. Rather, the only nearby gravitational field strong enough to visibly alter the path of light was the sun's. During the closing days of World War I, the British astronomer Arthur Stanley Eddington, an early advocate of relativity, sought to take advantage of an impending solar eclipse in 1919 to put Einstein's theory to the test. British expeditions traveled near the equator where the eclipse would be visible to determine whether the light from stars bent as it passed near the sun. If so, their positions would appear to have shifted. In a dramatic announcement at a joint meeting of the Royal Society and the Royal Astronomical Society, Eddington claimed to have found precisely the amount of bending Einstein had predicted and thus to have confirmed Einstein's gravitational equations. Even before 1919, some physicists and mathematicians, particularly in Germany, had become aware of Einstein's work on relativity.[15] But the bulk of Americans, both lay people and scientific researchers, remained largely unaware of it. J. S. Aimes, a professor at Johns Hopkins, confessed in 1920 that, though he had heard something about Einstein's 1918 paper and knew of several reviews of it, "I was one of those who had postponed any serious study of the subject until its immense importance was borne upon me by the results of the recent eclipse expedition."[16]

That such a crucial and dramatic test, appearing in the latest scientific journals, should have caught the attention of scientific researchers such as Aimes hardly seems surprising. But Aimes noted that he was first made aware of the significance of relativity and its confirmation "by the articles in newspapers."[17] The timing of Eddington's test more generally corresponded with a dramatic increase in the desire of many scientists and admirers to reassert the role of established science as an arbiter of knowledge in general American culture. Some popularizers were motivated by desire to justify and solidify gains in support made during the war in the name of national defense. Others saw science as the only remaining constant in a nation grappling with rapid change. To an even greater degree than in the late 1800s, postwar American civilization was fascinated by its own economic, social, technological, and political development. The so-called "jazz age"—a label not always used to denote something positive—included radios that pulled human voices from the air, skyscrapers that emerged from city skylines, and advances in flight that promised to make aircraft nearly as commonplace as automobiles.[18] Science sometimes became an important means of making sense of and successfully navigating the new world Americans saw emerging around them. Some scientists, for instance, portrayed science and scientific thinking as important aspects of American democracy and bulwarks against the spread of totalitarianism.[19] A variety of scientific sources of advice blossomed in the early twentieth century, including "scientific management," "scientific advertising," and "scientific motherhood."[20] Anthropologist Margaret Mead's *Coming of Age in Samoa* (1928) in particular offered what some readers thought might be the means of reconstructing American sexuality. By 1929, the *New Republic* noted that "popularizers of science for the magazines and for Sunday newspapers find their services in greater demand than ever before."[21]

The post–World War I drive to bring science to the general public was also a reaction against what many contemporaries perceived as a previous lack. As some late-nineteenth-century observers had feared, specialization among scientific practitioners steered them away from close contact with the popular sphere. As the golden age of popularization led by figures such as Tyndall and Huxley receded, the rapid pace of scientific discovery and the growing ability to make a living doing research turned many scientific practitioners inward. Members of the scientific community spent much of their time in institutions closed to general participation, dealing with information that was generated with little concern for general consumption. A 1906 columnist in the *Nation* lamented that "one may say not that the average cultivated man has given up on science, but that science has given up on him."[22] Beginning in the 1920s, the subtleties of research in many areas seemed to undercut the accessible advice offered by the turn-of-the-century, evangelical public health and eugenics crusades, and both lost momentum in their previous forms by the 1940s.[23] Over the first several decades of the twentieth century there was also a more general slump in the popularization of science in American magazines.[24] By 1915, the venerable *Popular Science Monthly*, though still monthly, was a little less popular, losing some $10,000 per year.[25]

But scientists were not alone as public arbiters of scientific knowledge. Journalists also played an important role in the postwar drive to spread scientific knowledge. In 1921, veteran newspaperman Edward Scripps formed Science Service, a syndicated news service dedicated to the nationwide distribution of science-related stories. E. E. Slossen took the helm of the new organization.[26] Science Service was a sign of things to come. During the early twentieth century, readers of general-audience magazines encountered a growing percentage of science articles written by journalists and professional popularizers who did not actively practice science.[27] Even though many people writing about scientific information were not themselves scientists, science journalism as an occupation was tightly linked to the scientific establishment. Science writers depended on scientific workers for their stories. Many of them also had backgrounds in various scientific disciplines.[28] Slossen, for instance, was originally trained in chemistry. At the same time, science writers began to distinguish themselves as a distinct professional group and a new source of at least reflected scientific expertise. As a sign of their growing sense of community, a number of popularizers gathered in 1932 to form the National Association of Science Writers (NASW).

Like the professionalization of scientific practice, the formalization of scientific communication in the mass media also involved the construction of more prominent boundaries around science. Though science journalists worked to bridge the perceived gap between scientists and other Americans, their very existence also deepened and more fully established it by institutionalizing the role of the "buffer." The formalization of popular science and the development of privileged and sanctioned sources of general-audience scientific information were also symbolic of broader trends toward the management of the expanding volume of information available in the United States. During the interwar years, for instance, the Book-of-the-Month Club began offering guidance to bewildered readers. Syndicated columnists multiplied in newspapers across the nation between 1925 and 1935, offering readers more personal perspectives and less anonymous guidance on the news of the day. More generally, commercial publishing firms increasingly standardized public sources of information. The booming advertising industry served as one of the most pervasive guides to contemporary culture, not only the explosion of consumer goods, but also attendant ideas about what life was like in modern civilization.[29] Between the World Wars, scientific popularization often became intertwined with advertising and the emerging occupation of public relations.[30]

Such an environment meant that, in order to stand out, scientific stories had to routinely be "jazzed up." Such was the judgment of Henry Crew, professor of physics at Northwestern University and chief of the division of basic sciences at the 1932 Centennial of Progress Exposition in New York.[31] Relativity could hardly have been jazzier. The relativistic universe of four dimensions in which objects moved to and fro at nearly the speed of light and behaved in ways that raised a variety of brain-teasing paradoxes seemed destined to capture attention. Likewise, its creator was also a topic of widespread curiosity. Einstein's visit to New York in 1921, the same year Science Service was formed, took place amidst large crowds of both onlookers and reporters. A number of high-profile tests of the theory also provided fodder for news stories. In 1923, a team of Canadians made a second set of eclipse observations, this time in Australia. During the same year, a Lick observatory expedition sought to verify the original eclipse data, an attempt that one commentator called the "greatest scientific test which has been made this century."[32] In 1925, Michelson repeated his famous 1887 ether-drift experiment. Judging by the coverage it received, his replication was an event of some importance. In 1925 and 1926, Dayton C. Miller, who had independently repeated the ether-drift experiment, also appeared widely in the pages of periodicals, even though his results were the opposite of Michelson's. Miller announced his detection of an ether wind, thus invalidating all of Einstein's ideas.[33]

As the Miller experiment indicates, relativity sometimes generated controversy, which only spurred media coverage. In 1921, both the *Catholic World* and Henry Ford's *Dearborn Independent* discussed the theories of engineer Arvid Reuterdahl, who claimed to have anticipated all of the famous German physicist's discoveries and to have done so in a much more God-centered and Christian-friendly framework than had the Jewish Einstein. Several other American critics also emerged, including the colorful and increasingly eccentric Thomas Jefferson Jackson See, an astronomer at the West Coast Naval Observatory, who attacked Einstein on a number of fronts, including mistakes in arithmetic. Most observers quickly dismissed See, but Columbia University's Charles Lane Poor presented a greater challenge. Poor rejected the 1919 eclipse observations because, he asserted, both the deflection of starlight and other gravitational anomalies seemingly explained by relativity could be more simply accounted for by assuming an envelope of matter distributed in some way around the sun. Poor and See appeared frequently in the news in 1923 and 1924.[34]

Such attention helped to feed popular interest. In one of the first biographies of Einstein, Alexander Moszkowski portrayed the intensity of the initial public reaction to the theory, in which "women lost sight of domestic worries and discovered co-ordinate systems, the principle of simultaneity, and negatively charged electrons." Though exaggerated, his claim echoed the fascination of many Americans with relativity. A librarian at the New York Public

Library reported a steady stream of ordinary people interested in relativistic ideas as early as 1920. These library patrons included "an elderly lady who seldom has an opportunity to visit the library" and who "spent three afternoons with us, 'having a riotous time' as she expressed it." Despite such enthusiasm, however, not all Americans were immediately fascinated by science. Just as the 1930 riot was not all it seemed, for instance, the intent of the crowds that gathered in New York in 1921 during Einstein's visit was not as transparent as many contemporary observers suggested. Large numbers gathered to welcome the noted Zionist Chaim Weizmann, who led the party that included Einstein among others, were misinterpreted by the English press as adulation for Einstein. Only later, when the Yiddish press noticed the great interest in Einstein among gentiles, did it pay him any sustained attention.[35]

Indeed, science was not always on the minds of all Americans. The removal of scientific practice and knowledge from broadly public culture presented barriers to widespread diffusion. Outside of specialized sources of news, such as Science Service, or science-focused magazines intended for wide audiences, such as *Science News Letter*, scientific knowledge had to compete with news from the worlds of sports, politics, and Hollywood in the growing marketplace of ideas. Most magazine editors approached science "as just one of many topics suitable for literate conversation, not necessarily deserving of special attention in every issue."[36] Discussion of science in the *New York Times*, traditionally a newspaper with high scientific content, amounted to less than 2 percent of its total coverage. In fact, the growing otherness and unfamiliarity of science sometimes made it a difficult sell. During the interwar years, despite concerted attempts at popularization, magazine articles about science appeared with less frequency than they had during most of the nineteenth century.[37] Nor were advertisers necessarily quick to turn to scientific appeals in their guidance to modern consumers. During the mid-1920s, only about 5 percent of advertisements in the ad-rich *Saturday Evening Post* appeared to invoke science in any way at all, and far fewer as a major draw. The vast majority of ads relied on such traditional, familiar, and easily accessible standbys as glowing personal testimonials, claims of attentive customer service, or aesthetically pleasing design.[38]

Indeed, a large part of early-twentieth-century American culture and rhetoric aimed at empowering citizen-consumers to make choices between competing facts, behaviors, and products. The Progressive movement of the first quarter of the century instituted a wide variety of such reforms, including the passage of the Seventeenth Amendment to the federal Constitution, dictating the direct election of senators in 1914; the Federal Trade Commission Act of the same year, which was designed to intensify the battle with corporate monopolies and trusts; and the achievement of women's suffrage with the adoption of the Nineteenth Amendment in 1920. Paeans to democracy were particularly intense during World War I, as Americans faced off against central European monarchies, and in the years leading up to World War II, as a counterpoint to the spread of fascism. Yet there was also a sort of paradox simmering inside celebrations of Americans' expanding freedom to choose. In science as well as elsewhere, the early 1900s also witnessed the development of social institutions and networks that sought to direct or even limit people's options. Progressive-era reformers, often themselves middle class and imbued with a healthy respect for professional training, typically juxtaposed concerns for popular empowerment with reliance on experts.[39] By the second quarter of the century, the duality between democratic and authoritarian impulses was nowhere more evident than in the mission of the advertising industry to both enable and control popular consumption. Still, whatever the underlying tensions of the emerging consumer culture, emphasis on the ability of information consumers to select freely from the expanding marketplace of ideas did open up space to sidestep science in favor of other kinds of knowledge.

While the jazziness of relativity helped it overcome such barriers, news about Einstein declined toward the end of the 1920s. On the eve of the 1930s, however, he was back in the headlines. The appearance of his unified field theory in 1929 provided fresh motivation for coverage, and many media sources translated this event into news. The *New York Evening Post* went to considerable trouble to print a "radiophoto" of the first page of Einstein's pamphlet on the subject.[40] But there was more to the second wave of Einstein's popularity than news events. Popular discussion tended to focus more on Einstein as a personality than previously. A 1929 article in the *Literary Digest* entitled "After All, Einstein Is a Human Being" announced the shift nicely. Another publication asked readers to imagine a violin contest between Einstein and Mussolini (for a short time, another celebrity of the period) rather than four-dimensional space-time.[41] All of this was consistent with changes in science popularization. The advent of the Great Depression, and fear over a general reaction against science and mechanization, further impelled others to highlight the progressive aspects of science, its contributions to American society and culture, and very frequently its human side.[42] More generally too, the hard economic times nudged many advertisers toward more humanized approaches. By the "early 30s advertisers had thoroughly accepted the idea of a public hunger for personalized communications, and new experiments in the 'personal touch' steadily proliferated."[43]

Meanwhile, Einstein himself was changing in ways that made him far more competitive in the contemporary media environment. Over the course of the 1920s, he had turned from a fairly sober-looking German scholar into a wild-haired, soulful character. He had also become more adept at handling the press and more willing to offer glimpses of his personal beliefs. In a 1921 conversation with a *New York Evening Post* correspondent, Einstein had awkwardly mentioned his dedication to pacifism. By 1930, however, stopping over in New York on the way to Caltech, he handled the mobs of reporters that gathered around him with smiles and a few jokes. In the same year, he wrote an article for *Forum* magazine entitled "What I Believe," and an extended statement of his on pacifism appeared in the *New York Times*.[44] Einstein's movement beyond the shadow of his equations helped discussion of him and his ideas spread into periodicals that previously had expressed little interest. Even the venerable *Nation*, which consistently displayed a lukewarm reaction to Einstein and his theories, roundly praised the physicist for his pacifism. The *Nation* had advocated pacifism for some time; a columnist wrote in 1930 that it was "refreshing to hear the world's most famous scientist say the same thing."[45] Not everyone was happy with his social views, however. Members of the Women Patriot's Corporation protested Einstein's attempt to secure a visa for travel to America on the basis of his supposedly leftist and radical ideas.[46] Nevertheless, the *Christian Century* eventually celebrated Einstein's eventual move to become a United States citizen, noting somewhat emotionally in 1940 that "Professor Einstein has been an exile in recent years, but he is an exile no longer. He is at home—an American in America."[47]

Science and Common Sense

More impermeable scientific boundaries had made room for a visible and robust scientific establishment on cultural maps. But they also heightened the risk that jazziness could tip over into a riot of boundary violation. The ways many people talked about Einstein and his relativistic physics during the 1920s and 1930s drew considerable attention to the barriers between science and the rest of contemporary culture. As we have seen, these barriers had some deep roots in the nineteenth century. During the late 1800s especially, the rhetorical construction of a gap between scientific and ordinary knowledge had helped to sever the activities of popularization and research, partially insulating researchers from what could be increasingly dismissed as "external" influences. It also justified the growing industry of popularization, underwriting the genre of popular science, and motivated popularizers to do their work. Had scientific conclusions about the world simply confirmed what people already knew or could get from other sources of information, there would have been less urgency in diffusing them widely. Even early-nineteenth-century phrenologists knew this, though they typically balanced claims of novelty with pointing out how phrenological science confirmed age-old truths. Depictions of the unprecedented newness of scientific discoveries were much more aggressive a century later in ways that not only reflected the greater independence of scientific researchers but the contemporary information landscape as well. To survive, newspapers and magazines needed large audiences to lure advertisers. As mass media made more and more knowledge available to Americans, stories in turn needed an edge to lure audiences.

If any early-twentieth-century scientific development seemed to contradict common wisdom, it was relativity. It was certainly an odd choice for those hoping to show the cultural relevance of science, given its highly technical and notoriously abstruse nature. Its implications and accoutrements, from shrinking measuring sticks and slowing clocks to the fourth dimension, gave even experienced scientists pause. Nevertheless, scientists, journalists, and popularizers worked hard throughout the 1920s and 1930s to bring relativity to the ordinary person. In doing so, they drew on and extended trends that were already a part of popular rhetoric. Early focus on scientific "breakthroughs" had appeared in coverage of such medical triumphs as Louis Pasteur's rabies vaccine in the 1880s. Over the next decade, an emphasis on the singular and stunning discovery became a normal part of newspaper depictions of big medical developments.[48] Americans talking about relativity used similarly dramatic, eye-catching language. "Lights All Askew in Heavens," blared a November 1919 headline in the *New York Times*. Einstein himself was the "Hero of the New Physics," the "wizard of four dimensions," or the "John L. Sullivan of Science" for having "knocked out Euclid and driven Newton into a corner."[49] Many of the works that used such animated rhetoric included what became a standard series of anecdotes and images that highlighted the incomprehensibility of Einstein's ideas and reaffirmed the unbridgeable boundaries that separated the average reader from the latest scientific knowledge. In the struggle to make relativity relevant, the very notion of public attention to science sometimes appeared to involve something riotous.

Depictions of the radical newness of relativity reached their zenith in claims about its "REVOLUTIONARY RESULTS," sometimes in terms almost as blaring as their script.[50] But novelty had its dangerous aspects too. Poor called relativity "bolshevism" in physics on several occasions. In 1922, on the heels of Einstein's visit to the United States, he elaborated on such claims by musing that "to a degree perhaps never equaled before, the social, religious, political, economic, and most other maxims that have served as guides in the past are receiving scant deference and are often suffering open questioning or active hostility."[51] Even advocates of relativistic physics sometimes sought to temper images of a revolutionary science. Einstein himself explained in his *Evolution of Physics* (1938) that "creating a new theory is not like destroying an old barn and erecting a skyscraper in its place." Instead, it was "rather like climbing a mountain, gaining new and wider views," and "discovering unexpected connections." The starting point of the trip did not cease to exist, "although it appears smaller and forms a tiny part of our broad view gained by the mastery of obstacles on our adventurous way up."[52] Despite a tendency to emphasize the sensational, media attention was not completely saturated by revolutionary rhetoric either. Numerous authors stressed that Einstein had built on old foundations rather than starting from scratch. A 1920 article in the *Literary Digest* portrayed Einstein as a merely the "Revisor" of the universe and noted that Einstein kept a bust of Newton in his office.[53]

On the whole, however, few popular accounts of relativity could resist at least a little revolutionary rhetoric or the implication of intellectual power that flowed from it. Nor did such declarations drop only from the pens of journalists. Much of it came from eminently quotable scientists, such as Max Planck and Werner Heisenberg. Even the distinguished Michelson noted at a 1931 dinner given for a visiting Einstein at MIT that his famous experiment had been conducted with "no conception of the tremendous consequences brought about by the great revolution which Dr. Einstein's theory of relativity has caused—a revolution in scientific thought unprecedented in the history of science."[54] The fundamental newness and unfamiliarity of relativity provided it with the potential of depicting things in a new and truer way. H. Horton Sheldon claimed that relativity fulfilled the very purpose of science in promising a "better conception of the place our earth holds" in the universe "and of the importance which we can attach to our own existence."[55] Likewise, the philosopher Hans Reichenbach claimed Einstein provided the "cosmic perspectives" needed to "be able to experience a feeling for our place in the world."[56] Public commentators often dipped into the past to show how this had happened before. Slossen ended one of his articles with a rare flourish by comparing contemporary sensations confronting relativity with those undoubtedly experienced by "those poor frightened people who lived in the times of Copernicus and Galileo when they were told that the solid earth was not supported by anything."[57] In time, what seemed odd would become natural as a new sense of reality emerged. Levi Gruber, an associate editor at *Bibliotheca Sacra*, noted that the new mindset contained in relativistic physics was no stranger than the implications of Copernicanism, which had become perfectly consistent with common sense.[58]

While such language helped to set relativistic science apart as special and powerful, it created boundaries that made its popularization nearly as much of a paradox as those involving moving trains and clocks regularly drawn from the theory. In the same way that late-nineteenth-century debate over evolution had helped to spread the notion of a conflict between science and religion rather than ignite it, discussion of relativity provided a vehicle for images of a potential clash between science and common sense. During the 1800s, as we have seen, the notion that science was simply organized common sense circulated widely, often helping to create publicly accessible images of scientific knowledge and practice. Commonsensical depictions of science still appeared during the early twentieth century, but much less frequently and often in the writings of those who themselves felt alienated by recent scientific developments. Poor characterized relativistic physics as "so bizarre it shocked common sense."[59] Opponents of the new physics argued that while it might be acceptable philosophy, abstract mathematics, or mysticism, it was not good science from the "commonsense point of view."[60] Some friends of relativity sought to show that Einstein's ideas, undistorted by journalistic zing, really were commonsensical. But others retorted by echoing Eddington's quip that "the cosmos is undoubtedly regulated by sense, but not always common sense."[61] A 1929 article in *Science Monthly* by a physicist from the State University of Iowa admitted that "common sense is the mother of science," but claimed that "as science became older, its laws became firmer and truer and stranger."[62]

The growing strangeness of certain areas suggested that, like science and religion, scientific knowledge and ordinary experience occupied distinctly separate spheres. As the nineteenth century progressed, the sphere of science had often been focused on the study of the physical world. But the narrowing of scientific ground did not stop there. Twentieth-century science sometimes appeared to involve only parts of the physical world, particularly those aspects inaccessible to the unaided senses, spending its time out among the galaxies or deep within the atom. At other times, it seemed to be bound for more distant destinations, led by the twelve (or three or four) men who had reportedly been able to understand relativistic ideas. Gleeful authors compared the world of relativity with that of Alice's Wonderland or the territories of Gulliver's travels. No one urged science on its way to stranger realms quite as much as Eddington. "The frank realization that physical science is concerned with a world of shadows," he wrote, "is one of the most significant of recent advances."[63] Elsewhere, Eddington contrasted the world of science and ordinary experience through a comparison of the "scientific" and "familiar" table on which he wrote. The latter was a well-known and solid object. The former was a "a collection of electrons moving with prodigious speeds in empty spaces which relative to electronic dimensions are as wide as the spaces between the planets in the solar system."[64] The contrast between these two representations convinced at least one observer that if science and the ordinary did not exist in separate spheres, they should. If Einstein were correct, "then smash go our conceptions of a universe of flowers and gold and singing birds and stinging bees, of which poets have written for centuries, and we shall have to recon-struct our ideas of the cosmogony of space and worlds . . . according to a fixed rule and line, which, incidentally, takes count of time, rather than eternity, and of the beginning as well as the ending of things."[65] So powerful and so potentially inhuman a science, cut off from the ordinary world, could take on an edge of danger in more mundane ways, too, as it did amidst early-century concern over human experimentation and vivisection behind closed labora-tory doors and poison gas on the World War I battlefield.[66]

But changes in science were not alone responsible for the growing bound-ary between science and the rest of human experience. The remoteness of sci-ence was further exacerbated by a growing tendency among many intellectuals and journalists to depict the general public as largely incapable of dealing with complex ideas in a conscious way. As early as 1922, Walter Lippmann "had begun to knock the 'public' off the perch that the rhetoric of democracy had built for it."[67] The picture of an irrational public, influenced by the ideas of Sigmund Freud and the supposed success of World War I propaganda, was a far cry from the beliefs of the Progressives. Muckrakers such as Lincoln Stephens and Ida Tarbell sought simply to provide their readers with data and expected that sober deliberation would finally motivate the masses to demand reform. The advertising industry deeply imbibed the notion that it could manipu-late an essentially passive and primarily emotional public. Advertising copy, which had once listed the specific virtues and characteristics of a given prod-uct, began to associate goods with lifestyles and attitudes. Such developments resulted in both a "disillusioned, distinctly modern attitude toward facts" in journalism and eventually, in reaction, the first steps toward an explicit creed of "objectivity."[68]

In the shadow of the boundary between science and the ordinary world, interest in relativity among nonscientific audiences became a kind of aber-ration, as if ordinary people could not seek to understand relativity without causing a riot. The Nation, in 1929, classified fascination with Einstein in the same category with the "annual sea serpent, the seven-year mutation of our bodies, the jargon of Freud, the messages from Mars, the latest gossip concern-ing the health of the President, and the prophesied end of the world (which has again just failed to occur according to the prophecy)."[69] In light of con-temporary psychological theory, the very incomprehensibility of Einstein's sci-ence sometimes appeared itself as a sufficient explanation of popular attempts to understand it. According to an array of commentators during the 1930s, including at least one psychoanalyst, popular attention to Einstein arose precisely because he was not intelligible to the average kinds of people "who do not, and never can hope to, understand his work" and yet "idolize him as they idolize Babe Ruth."[70] When a page of Einstein's pamphlet on the unified field theory appeared in the New York *Evening Post*, a columnist in the *New Republic* remarked that "their readers probably enjoyed the subtle compliment of the assumption that they might be able to comprehend these complicated formulae" and suggested that "while they know very well that there is an aristocracy of intellect, probably they like to pretend, even with an ostentatious self-depreciation, that there is none, that one man's brain is as good as another's."[71]

Facts and Theories

As the content of the latest science seemed to recede from public understanding, the notion of a scientific method became increasingly important to the work of many popularizers seeking to clarify the boundaries of science for the average reader. But here too, those discussing relativity in magazines and books demonstrated how easy it was to run afoul of those boundaries in the quest to make scientific knowledge relevant. For some time, talk about method had played a prominent role in expressing the nature of science and the ways in which its knowledge compared with other statements about the world. During the early 1800s, as we have seen, most Americans writing about science stressed the unity of truths and so science shared its processes with a wide array of other fields. By the later nineteenth century, methodology had increasingly become a means of distinguishing science from other realms, a tactic that culminated in references to a uniquely "scientific method." Invocation of scientific method reached new and unprecedented heights after 1900, partly a testament to the spreading perception of science as a well-bounded, separate field.[72] Among those who wanted to be sheltered within such boundaries, it became increasingly important to show that one's methods were not only sound but also scientific. Much of the usage of scientific method during the 1920s emerged from the social sciences, which had often seemed to be beyond the pale in the midst of strongly physical definitions of science.

Scientific method also provided a commonly used means of demonstrating the power of science beyond strictly scientific boundaries. References migrated to such diverse realms as law, parenting, journalism, even music instruction and the raising of bullfrogs. In books such as Eby's *Complete Scientific Method for Saxophone* (1922), Martin Henry Fenton's *Scientific Method of Raising Jumbo Bullfrogs* (1932), and Arnold Ehret's *A Scientific Method of Eating*

Your Way to Health (1924) authors used the scientific method to sell new ways of solving old problems. Those eager to endow the masses with an appreciation for science proclaimed most loudly that the "greatest gift of science" was "the scientific method."[73] In a 1932 address to journalists in Washington, DC, physicist Robert Millikan told his audience that the "main thing that the popularization of science can contribute to the progress of the world consists in the spreading of a knowledge of the method of science to the man in the street" and showing how it could be used to solve the problems of life.[74] During the 1930s, as fascism spread abroad, some admirers of science and defenders of democracy equated a scientific attitude specifically with democratic values and the mindset of the ideal citizen.[75] Perhaps no group promoted scientific methodology with as much relish as educators. Speaking to the educational section of the American Association for the Advancement of Science in 1910, John Dewey charged that science "has been taught too much as an accumulation of ready-made material with which students are to be made familiar and not enough as a method of thinking." By 1947, the *47th Yearbook of the National Society for the Study of Education* declared that there "have been few points in educational discussions on which there has been greater agreement than that of the desirability of teaching the scientific method."[76]

The idea of a uniquely scientific methodology opened the possibility of reestablishing the harmony of truth under the banner of science, while retaining images of a well-bounded, independent, and thoroughly professional scientific enterprise. Scholars have sometimes depicted what is often called hypothetico-deductivism—the proposal, then rigorous testing, of hypotheses, and the modification or rejection of those that proved unfit—as the "orthodoxy of the 20th century" in place of earlier and looser Baconian or inductive notions.[77] The 1919 test of relativistic predictions provided an example of hypothetico-deductivist verification writ large. Additional tests provided similar examples. But explicit hypothetico-deductivist rhetoric existed almost nowhere but in classrooms. The range of methodological ideas revealed by discussion of relativity varied between far greater extremes than even the relative chaos of late-nineteenth-century debate. On the one end existed the traditional invocation of facts, even in ways that sounded like rehashed induction. For Millikan, scientific method started "by trying to get to definite facts, and then allows these facts themselves, with their inevitable consequences, to determine the direction in which conclusions are formed."[78] In a similarly inductive vein, astronomer Henry Norris Russell claimed that Albert Michelson and Edward Morley's repeated failure to detect the motion of the earth through the ether in 1887 "led by degrees to the development" of relativity theory. Other commentators cited "several decades" of experiments, those of Michelson-Morley among them, from which relativity grew, though Poor criticized the tendency to inflate the "six meager observations of Michelson" into "a long series of observations made at many places and at many times of the year."[79]

Such accounts of methodology did little to set science apart as unique or special, because anyone could look at facts. Instead, as we have seen time and again, it was the kind of facts involved that helped to define science. In particular, the raw materials of scientific method increasingly became measurements, preferably with large or obscure machines. This view reached its ultimate development in Percy Bridgeman's operationalism. Inspired by Einstein's willingness to critique such fundamental notions as space and time, Bridgeman, an experimental physicist, argued that entities should be defined by the operations that detected them. Unmeasurables, such as the ether or absolute time, had no place in science. Bridgeman offered a basically inductive view of scientific method, seeing it as the result of attention to facts. His "pure empiricism" required the intensification of the age-old belief "that the *fact* has always been for the physicist the one ultimate thing from which there is no appeal, and in the face of which the only possible attitude is a humility almost religious." Recent science, including relativity, was not the result of a "swing of the fashion of thought toward metaphysics" but rather "forced upon us by a rapidly increasing array of cold experimental facts."[80] In many ways Bridgeman's views paralleled some of those of Eddington, who sometimes talked about basing science on "pointer readings." Psychological behaviorists advocated a similar method for studying humans.[81] And while neither operationalism nor behaviorism dominated popular discussion, the basic idea that science depended on measurements with equipment was a common one. A 1924 article in *Science Monthly* stated bluntly that science did not involve explaining "the obvious facts of everyday experience," but rather worked with facts "discovered by the more searching processes of the laboratory and observatory."[82] Several years later, the biologist David Starr Jordan asserted that "science is human experience tested and verified by our instruments."[83]

Such depictions helped to show the basic difference between the way science operated and the way ordinary people approached the world. It also helped to support claims that the essence of the scientific method consisted in its objectivity, its detachment from the familiar world of emotion and prejudice. Einstein made no secret of his own desire to transcend the "merely personal" realm through his science. Journalists, who looked increasingly to the method of science as a guide in their own work, saw the essence of that method in its objectivity.[84] Methodological views centered on machines and measurement also coincided well with images of the laboratory that proliferated widely during the early twentieth century. Invocation of the laboratory may have had its greatest impact on images of medicine and medical researchers, but its influence did not stop there. In 1887, Joseph LeConte specifically denied experiment, typical in chemistry and physics, a place in the study of life because biology was simply too complex. Seven years later he contrasted the old and new natural history by placing the former in the field and the latter in the lab. By 1916, John Merle Coulter, a biologist at the University of Chicago, had divided the history of biology into three phases: speculative, observational, and experimental. Observation allowed probable inferences but only laboratory work could demonstrate beyond doubt. The decade around the turn of the century witnessed similar growth in lab rhetoric among psychical researchers and social scientists, along with greater emphasis on repeatability and quantization. Sometimes scientists themselves became a reflection of their machines.[85]

A machine-based view came at the cost of putting the methods of science in a restricted light, unable to transcend its "pointer readings" or statistical tables of measurables. Relativity itself showed how limited human perception of the universe could be. Those hoping for a more expansive vision of science and its role in the world preferred a methodology with more power and scope. American scholar Howard Roelofs recalled much confusion in his youth over the precise nature of the scientific method, but he knew that it worked "miracles," something mere calculation could never do. Instead, it involved intelligence and imagination, elements that could not be taught "by a year's course in a laboratory science," where "all too often only measuring and counting are learned." For others, scientific method relied on "insight" and on thinking about events "until they become luminous."[86] Einstein himself noted "how much in error are those who believe that theory comes inductively from experience."[87] The diffusion of scientific method as a mindset rather than a technique, and sometimes associated with virtues such as honesty and healthy skepticism, sometimes converged with concern over the spread of totalitarianism in the late 1930s and 1940s. Scientific method provided a balance between an open and a critical mind, which allowed citizens to evaluate claims made by political figures. Its application by the American public foreshadowed a true "science of democracy."[88]

Theory provided an important keyword for twentieth-century depictions of an imaginative scientific method, though there was little agreement on what theorizing meant. For those sympathetic to experimental and practical views of science, theories were merely tools. Slossen, in his 1920 popularization of relativity, claimed that most laymen really did not understand "how firmly" the scientist held to facts and "how loosely" to theories.[89] Bridgeman also claimed that theories could be useful tools. But he hesitated ever to call one true, since experimental tests always contained some error.[90] Others, more sympathetic to an abstract view of science, stressed factors independent of facts, such as the convenience and simplicity of theories, often citing the French philosopher Henri Poincaré as their guide. Robert Carmichael, a mathematician at Indiana University, suggested that one should accept relativity, even if it had not been conclusively demonstrated, "provided it furnishes him with the most convenient means of representing external phenomena." James Jeans brought out Occam's razor—the notion that the simplest explanations should be preferred and a tool rarely discussed during the nineteenth century—in his *Mysterious Universe* (1930). Aesthetic factors, too, began to figure more prominently. A columnist in *Time* noted that physicists spoke of Maxwell's equations "as if they were a beautiful painting hung in a museum."[91]

Visions of science directed by such an idealistic method could reach almost anywhere, but that was one of their major weaknesses. It seemed constantly poised to violate the boundaries between science and other realms. Karl Pearson had written in 1892 that "all great scientists have, in a certain sense, been great artists; the man with no great imagination may collect facts, but he cannon make great discoveries," and in many accounts, Einstein appeared as the essential Pearsonian scientist, transcending mere facts by the power of his creative genius. The method of the German physicist was "surprisingly analogous to that of the artist," who created "a clear, harmonious world of thought." From such thought and "a very few observations," Einstein was able to "divine how the cosmos was made."[92] But not everyone approached such border violations with so much calm. Poor criticized the German's use of "artistic license" and claimed that the foundation of his general theory was merely a "version of Jules Verne's story of an imaginary trip to the moon." Under the unfortunate influence of relativity, another critic noted, the "almost tangible ether gives way to the Frankenstein of modern physics," four-dimensional space-time. A columnist in the *Nation* complained that "philosophers who could not solve a simple differential equation are beginning to assert that the mathematical treatment of the fourth dimension or the intricacies of the Quantum Theory furnish proof of the existence of God."[93] Einstein's image appeared in less savory cases of transgression, too. In 1932, when a California psychic claimed Einstein's endorsement, one Hollywood resident who believed that Einstein had in fact willingly lent his good name to the psychic exclaimed, "And so the scientific method goes crashing to the ground."[94]

Engineers and Prophets

As scientific knowledge itself seemed to become more remote, method became an increasingly important resource for constructing boundaries around science and for making sense of its influence beyond those boundaries. Depictions of scientific practitioners as "unassailable experts" also loomed large in twentieth-century science talk.[95] Like method, such depictions typically served both to help demarcate science from other areas of knowledge and practice and as potential bridges for the extension of scientific power and authority. Among the many indications of more bounded images of scientific work was the prominent, though hardly universal, tendency in popular forums to associate practitioners of science with their institutions, whether observatories or universities. The spread of the term "scientist" also provided a key rhetorical tool in constructing borders around the community of scientific practitioners. During the later nineteenth century, usage of the label remained sporadic, as did the tendency to surrender science to a particular group of practitioners. Over the early decades of the twentieth century, invocation of the scientist in magazine article titles increased more or less steadily, and the word quickly became a staple in American discussions of science.[96]

Einstein in particular became an icon in American popular culture as the archetypal scientist, perhaps more than any other scientific practitioner before or since. His outspokenness on political and moral topics also seemed to invest the idea of the scientist with newfound power and relevance. More generally, historian Marcel LaFollette maintained that "the image of scientists as unassailable experts generally drew little questioning" during the early 1900s, and "this unquestioning acceptance extended even when they were speaking outside their areas of expertise."[97] But Einstein's forays into subjects other than physics often carried with them the sense of crossed boundaries. In characterizations and descriptions, his relevance, importance, and sometimes his missteps often followed not from his position as a scientist, but because he was something more. Again, as in the case of method, the separation from the practices of ordinary life that formed the basis of scientists' special position sometimes fostered a sense of remoteness and disconnection. At times, this disconnection called into question the ability of scientific practitioners to reach beyond the boundaries they came to embody.

As the boundaries around the scientific community grew more visible, the ability and willingness of self-perceived outsiders to comment on scientific matters decreased. Unlike Americans who wrangled over evolution, few of those who presented themselves as laymen offered any public criticism of relativity or its fundamental correctness. This separation did not occur without some tension. In light of contradictory tests of the truth of relativity, a 1925 article in *Outlook* jokingly asked, "Should the public allow a mere handful—an insignificant minority—of specialists, not more in fact, than one in 200,000 of our population, to dictate what we shall believe? Should we not pass a law forbidding the Einstein theory to be true—or untrue?"[98] The mathematical and technical intricacies of the new physics kept many critics at bay. Grassroots opposition to other kinds of scientific ideas, such as the antievolution movement, was still possible, as was the rhetoric that helped to justify it. Noted public opponents of evolution such as William Jennings Bryan continued to carry with them images of an open and communal science, rooted in a more traditional nineteenth-century scheme of science talk.[99] There was some popular resistance to relativity, too. But as scientific knowledge generally became increasingly complex and difficult for those without training, criticism from nonspecialists fell ever more into the hands of people outside the mainstream of public discussion. During the mid-1920s, James W. Cromer of Los Angeles, an amateur philosopher, took the trouble to self-publish two books disputing relativity, the first garrulously subtitled *A Scientific Criticism of the Einstein Theory of Astronomy, also with Sarcastic Remarks, Fun and Jokes* (1925).[100]

In contrast to such lightheartedness, scientists were often models of seriousness, caution, and neutrality, whose detachment guaranteed the reliability of their investigations. These depictions closely fit the Progressive-era image of the scientist. Historian Ronald Tobey claimed that when Progressive reformers talked about the scientific expert they had in mind something much closer to what modern observers might call an engineer than a researcher.[101] Such figures did not seek transcendence or offer enlightenment through a grand set of truths about the world. Their knowledge became the basis for technical and practical solutions to concrete problems. Their expertise opened the promise of control. Such images survived well into the 1920s. In a 1922 *Science Monthly* article entitled "How the Chemist Moves the World," Slossen depicted the chemist as "not merely a manipulator of molecules; he is a manager of mankind."[102] Early media coverage of Einstein sometimes depicted him in a very aloof light as well, more akin to a technician standing above the world than a passionate searcher after transcendent truth. One early account of relativity theory for a wide audience depicted him as working in his "laboratory in Berlin" and as being unconcerned with the events of World War I.[103] A 1926 article in the *Scientific American* emphasized his "sportsmanlike spirit" in the face of contradictory evidence to his ideas offered by Dayton C. Miller, another model of objective detachment. The true scientist was searching "only for the facts rather than a vindication of his own life work."[104]

Images of Einstein took on a distinctly less restrained tone during the 1930s, though depictions of many of the other scientists that surrounded the German physicist and his work remained somewhat conservative. A 1925 article in *Outlook* depicted most scientists as cautiously neutral on the question of the truth of relativity theory, "calmly sitting back for another ten or twenty years to wait and see more facts."[105] A drawing accompanying an article on Einstein in a 1921 issue of *Illustrated World* showed an expressionless man with short hair, graying at the sides, glasses, in a suit, surrounded by a plethora of equipment, including a microscope, telescope, and test tubes. Such an image seemed a stark contrast with the photo of a moderately wild-haired man that appeared beneath it.[106] Likewise, a 1936 photo in *Science Monthly* showed strongly contrasting characters in a rumpled, messy-haired, smiling Einstein shaking hands with the well-groomed, neatly dressed, short-haired, and expressionless physicist and president of the Board of Trustees of the New York Museum of Science and Industry, Frank Jewett.[107] Some of the harshest critics of relativity and the new physics often appeared as representatives of this more practical and aloof vision of science. Arvid Reuterdahl, himself the dean of the Department of Engineering and Architecture at the College of St. Thomas, cited the work of Charles F. Brush, a Cleveland-based electrical engineer and scientist, against Einstein. And a 1929 *Review of Reviews* article described Poor as a "yachtsman, retired business man, and professor of celestial mechanics in Columbia University."[108] Sometimes, the truly detached and neutral scientist even lost the search for truth in the glare of utility. In his 1920 *Easy Lessons in Einstein*, Slossen informed his potentially bewildered readers that they should recognize that "the scientist never bothers his head with the question of whether a particular theory is true or false." He only cared whether it, like any good tool, worked.[109]

By contrast, Einstein's emergence as a public figure on matters from science to morality, God, and pacifism in the 1930s was frequently accompanied by claims that he was something more than a scientist. A reporter who visited him in Princeton in 1934, after his permanent relocation to the United States., noted that his house "might have been the dwelling of a gloomy poet."[110] Likewise, a 1932 popularization of his scientific ideas claimed that "Einstein is also a musician. His pleasure in this hobby ranks with his pipe, his boat, and his work. . . . Perhaps the concert stage was robbed that we might have such a great creative thinker."[111] A 1929 article in *Review of Reviews* suggested that he was a "really a musician masquerading as a scientist."[112] At other times, he seemed to blend the qualities of the devotee of nature and the religious figure. A 1935 article on his commitment to pacifism depicted him as "more than the savant who has spent his life in the physics laboratories of Zurich and Berlin, more than the gentle lover of the violin and nature; he is a reformer who hates war with the fervor of an Old Testament prophet." Indeed, rather than calling him a scientist, it often seemed more appropriate to classify him with other great men or even geniuses, who often united "the qualities of both the artist and the scientist."[113]

In some ways, images of Einstein during the 1930s seemed to hearken back to an early-nineteenth-century view of the man of science, whose attempts to discover truths about nature were of a piece with the discovery and display of moral, social, and cultural truths. But such an option was far more difficult to maintain or even talk about in the 1900s, given the tendency to set those who practiced science apart. Einstein could not be an ordinary person, and he could not be a narrow scientist, lost in a world of detail, confined to the laboratory, or myopically focused on a specialization. With nowhere left to expand, depictions of him extended into the otherworldly realm of deep thoughts. While magazine articles about Einstein often sought to stress his personal and even ordinary characteristics in the second wave of his fame, showing him at work and at play, even enjoying an ice cream cone, these attempts were almost constantly undercut by depictions that portrayed him as a kind of higher being. Psychic Gene Davis, who later claimed Einstein's endorsement, recalled that his aura was "pure blue electric sparks, instead of color. It was just like talking to God."[114]

More mainstream sources also described something a little unworldly about the German physicist. A columnist in the *Christian Century* was pleased and seemingly a little surprised to find Einstein, "a great scientist who might be supposed to have his eyes focused on nothing nearer than Alpha Centauri, calling with the fervor of an evangelist for a decision day on the personal renunciation of war."[115] Authors repeatedly talked about Einstein as living in a world of his own, working locked in his mind or his attic, his only real link to the world being his second wife Elsa. A 1934 article in *Newsweek* described "a frightened-looking, pixie-like little man with a strapping hausfrau bodyguard. Pedestrians along Center St. looked at the little man's fuzzy graying hair and his black slouch hat. Then they looked again. It was Albert Einstein."[116] Slowly, this combination of transcendence and otherworldliness coalesced around the notion of the "theoretical physicist." A 1920 *Science Monthly* article noted that "next to the philosopher nobody so insistently tries to reach the 'ultimate' as the theoretical physicist."[117]

Alternately, just as in the late nineteenth century, criticism of the supposedly nonscientific ideas of Einstein and other scientists was sometimes couched in terms of limiting them to the proper, and seemingly narrow, realm of the scientist. In response to an article that highlighted Einstein's Jewish roots and their influence on his religious ideas, one resident of North Providence, Rhode Island, wrote that "in view of Einstein's imponderable contribution to scientific discovery, any attempt to localize him as a Jew or as a representative of the Jews is insulting and misleading. I hope that some day he will speak in your pages as a scientist, and purely as a scientist."[118] Einstein's identification with Judaism bothered some even before the 1930s. In discussing the 1921 visit to New

York, Slossen criticized Einstein's public association with Zionism, suggesting that scientists should adopt a more international perspective. A concerned citizen, responding to Einstein's supposed endorsement of the psychic Davis, wrote sadly, "Instead of coming to the support of what is sane and rational *consistently and always*, in harmony with his position as a great scientist, he has here made a tremendous and, in all probability resounding contribution to the success of superstition." It seemed, he continued, that scientists, particularly physicists, were "hopeless" once outside their specialties.[119]

Though readers of magazine articles and popularizations of relativity could have deferred to Einstein's opinions about the world beyond science, they had ample resources to justify resisting his social, political, and religious views. It was all too easy for scientific expertise to be overwhelmed by a variety of other factors, such as Einstein's suspect and leftist political commitments, which so upset the members of the Women's Patriot Corporation. On the other end of the political spectrum, American pacifists, who took comfort in Einstein's outspoken support for their cause, were able to dismiss him when they felt that his support wavered. An article in the *Christian Century* explained that the "Einstein creed for the ending of war proves to be as simple as his mathematical ideas and physical formulas are complex." Its author further hoped that his defense of pacifist beliefs would "cause many otherwise unapproachable minds to regard the program of absolute personal pacifism with new respect." Several years later, when Einstein mitigated his previously uncompromising pacifism in light of Nazi aggression in Europe, another *Christian Century* columnist reflected that "it appears that violence and non-resistance also are subject to the principle of relativity" and lamented that "absolute pacifism should be made of sterner stuff."[120]

The Religion of Science

During the nineteenth century, the links between natural and divine knowledge provided an archetypal example of the connectedness of science and its harmony with a broad range of truths about the world. In the twentieth century, too, a valuable and broadly relevant science as "the foremost expression of our culture and civilization" sometimes found itself cast in religious language. An opponent of relativity explained his opposition by noting that it was "the sacred duty of those endowed with the gift of understanding to protect Science above all things, by a religiously honest attitude toward it."[121] Supporters of Einstein often dipped into their religious vocabularies as well in their search to make sense of relativity. In fact, during the 1920s and 1930s, few areas of science attracted more religious rhetoric than the new physics, particularly the work of Eddington, Jeans, and even Einstein himself. But, as with other attempts to reconnect science to the rest of American culture, such depictions often came at the cost of violating rhetorical boundaries around the scientific and in the end frequently appeared to stray into the occult or simply strange rather than the profound or even relevant.

As Einstein became an important cultural icon, he attracted a wide array of religious references, not least because he himself wrote widely about his own views on religion and its relationship with modern science. A 1930 article in *American Magazine* depicted Einstein as "a mysterious figure, writing bewildering signs, uttering phrases beyond understanding—a high priest of science."[122] Einstein was not the only scientist depicted in religious terms. An admirer of Eddington, a devout Quaker, portrayed the British astronomer as not only a scientist "but a Christian mystic who interprets the philosophic significance and defends the religious implications of his austere formulas."[123] In fact, Einstein was no friend of orthodox religious ideas. He openly criticized the idea of a personal, anthropomorphized God. But he never couched such criticism in larger claims about innate tension between scientific and religious views of the world. In a conversation among Irish author James Murphy, widely known science popularizer J.W.N. Sullivan, and Einstein printed in a 1930 issue of *Forum*, James Murphy noted calls of some for "science to give a new definition of God." Einstein responded, "Quite ridiculous!" Instead, the German physicist claimed "that modern scientific theory is tending toward a sort of transcendental synthesis in which the scientific mind will work in harmony with man's religious instincts and sense of beauty."[124]

Such statements encouraged devotees of the new physics to hope that relativity and quantum mechanics might herald a new era of harmony between science and religion.[125] It became commonplace to contrast the new view of the universe offered by the latest physics with the simplistic mechanical models of "nineteenth century science." The discovery of the increase of the mass of the electron, one MIT professor noted in 1920, "suggested very strongly that the day for a mechanical explanation of nature has definitely passed."[126] In his *Mysterious Universe*, Jeans pointed to the tendency of all modern physics not only to deny materialism, but also to "reduce the whole universe to a world of light, potential or existent, so that the whole story of its creation can be told with perfect accuracy and completeness in the six words: 'God said, 'Let there be light.'" In support, Jeans cited the work of Robert Millikan, who sought to show the providence of God in the study of cosmic rays and who became a very public advocate for the harmony of science and religion. Most important of all, the new physics seemed to make room for the power of mind in the world. In its increasing appeal to mathematics rather than mechanics to explain phenomena, Jeans and others claimed that science had begun to depict a universe "more like a great thought than a great machine."[127]

Such harmony was as much a boon for science as religion. The bulk of the religious references used in discussing the new physics were less about replacing religion than about turning to religious language to express what seemed to be beyond ordinary scientific terms. Often, the combination of scientific and religious rhetoric involved movements toward expanded depictions of the nature and value of science. In a written communication to a Conference on Science, Philosophy, and Religion at the Jewish Seminary of America, Einstein reaffirmed the now age-old view that true science and true religion could never conflict, but he refused to completely separate them. Science, he claimed, "can only be created by those who are thoroughly imbued with the aspiration towards truth and understanding. This source of feeling, however, springs from the sphere of religion."[128] Repeatedly, advocates of relativity saw one of its primary values in expanding one's vision so as to reveal "a universe in which mind and spirit can at once be immanent and still transcend the fettering limitations of the gross environment."[129] One admirer frankly depicted relativity as an attempt to expand perception of the nature of science and "penetrate beyond the limits of experience." Ultimately, the author exulted, "Science at last appears to venture into the mystic fields of metaphysics," confirming that "matter is reducible to force, or, if you please, spirit."[130] Some enthusiasts such as Eddington seemed prepared to erase the boundary between the natural and supernatural altogether, a move inspired by quantum mechanics' abandonment of determinism. He argued that the new physics made it impossible "to stigmatize certain views as *unscientific* superstition," though they could still be rejected as *"bad science"* if circumstances warranted.[131]

The use of religious language to present expanded visions of the scientific may have helped to make science appeal to many readers. Einstein has as frequently been associated with his musings about God not playing dice as with $E=mc^2$ or the atomic bomb. But though such wanton mixture of scientific and religious rhetoric was generally acceptable in the early nineteenth century, by the 1920s and 1930s it occurred within a context shaped by discussion of science and religion as largely separate. A correspondent for the *New Republic* claimed that Einstein had become "our most recent Deus ex Machina (and what a Machina! When all straight lines are parallel thingamijigses are guaranteed to do the exact opposite from what they have always done)" and that "he is—or may be—no, he positively is the Savior, who by rubbing the magic stone will cure us of all our ills."[132] Einstein sometimes found his own religious pronouncements attacked as "singularly illogical and vague." One author noted that if "a theologian should write on religion and science with the combination of pontifical certainty and naive incompetence shown by Professor Einstein . . . scientists might well receive his utterances with more amazement than respect."[133]

Indeed, the "well-known conflicts between religion and science" had taken on new life during the 1920s. The notorious Scopes trial in Dayton, Tennessee, garnered enormous attention and inspired one self-styled champion of science, Maynard Shipley, to write a kind of modern update of what White and Draper called *The War on Modern Science* (1927). From the side of Fundamentalists, of course, the war was being waged on modern religion.[134] In popular rhetoric generally, the combination "science and religion" reached new heights of use in the attempt to make sense of such developments during the early 1900s.[135] From some perspectives, just as during the late 1800s, a science distinct from religion could also be privileged over religion. An article entitled "Priests in the Religion of Science" in a 1923 issue of *Review of Reviews* detailed the views of biologist Edward M. East, who reportedly believed that the "one means of solving all human problems" was "more science and still more science." He predicted that "the spiritual nourishment once given man by his church will be replaced with the physical nourishment offered him by science."[136] Sounding somewhat Tyndallesque, James Murphy also suggested that the work of Einstein and his physicist colleagues was the first step toward a "veritable cosmology, whose doctrines are tested by factual investigation of an impersonal kind." He wondered to what degree the new physics could be "turned into a philosophy which may go some way toward establishing practical ideas of life on the ruins of religions ideas."[137] Journalist Frederick Lewis Allen suspected that the popularity of science he claimed to witness during the 1920s was related strongly to a supposedly coincident decline in religion.[138]

Perhaps because of such tension, the great deal of talk about science proceeded without any religious allusions, particularly in areas of well-established scientific knowledge. The scientific display at the 1933 World's Fair, initially named the "Temple of Science," was later demoted to a "Hall."[139] Rather, by the years after World War I, most religious references in the discussions of science occurred at its frontiers, in the most extreme and seemingly abstract areas, such as cosmic rays or relativistic speeds. Within relativity it often hovered most noticeably around the idea of a fourth dimension and a non-Euclidean geometry. In 1932, H. L. Mencken lampooned the "New Theology" that had followed on the heels of the new physics "with a mathematical God compounded of straight curves and curved straight lines." Even Einstein warned readers that the "non-mathematician is seized by a mystical shuddering when he hears of four-dimensional things, by a feeling not unlike that awakened by thoughts of the occult." He was right. From the late nineteenth century into the early twentieth, the fourth dimension repeatedly appeared in the works of Spiritualists and occultists. In the wake of popular discussions of relativity, some of them returned to it with new vigor. The noted occultist P. D. Ouspensky, among others, discussed the possibility of expanding one's mind into the level of a "fourth dimensional consciousness."[140] Such excursions into metaphysics appeared to confirm the conclusions, as one Theosophist noted, of Madame Blavatsky.[141] Increasingly, the religion of science was the possession of those on the margins of American culture.

Ultimately, some supporters of relativity found themselves back-peddling from what they saw as the overenthusiastic, semireligious claims of fellow admirers of Einstein. Max Talmey, who had known Einstein as a child and emerged as a staunch supporter of relativity, criticized the speculations of some theoretical physics as "occult" or "cabbalistic" and claimed that relativity had nothing to do with religion.[142] Talmey was not alone in his assessment. The rhetorical boundary between science and religion remained strong enough to be a useful weapon in the hands of those opposed to all or part of relativistic physics. A critic of Einstein's ideas in 1936 attributed their popularity to relativity's ability to awaken "that sweet tickling feeling that the very mysterious and very mystical induce in" ordinary people.[143] Another skeptic claimed that relativity was not science but simply a "kind of new religion with Einstein for its prophet."[144] Mixing religion and science seemed almost to guarantee the marginalization of those who dared do it, either as a Fundamentalist reactionary or as a bleary-eyed mystic. For many other public commentators, as well, the religion of science offered by the likes of Eddington and Jeans seemed to constantly intrude into the mystical, occult, or odd.

The Limits of Science

Authors of early-twentieth-century magazines and popular books almost universally depicted science as distinct from other kinds of knowledge. Popular science talk had crystallized around a number of keywords and along several crucial boundary lines between science and everything else. Even if the precise position of those boundaries remained a matter of debate and sometimes confusion, their existence gained assent nearly without question. Science and religion clearly occupied different realms for the vast majority of public commentators. Those who practiced science were unquestionably a group apart. And though few people could agree about its nature, there was widespread talk suggesting that science possessed a method of its own. Just as in the late nineteenth century, the distinctions between science and other categories provided numerous means of expressing claims about its power and influence. The ideas of science, such as he-liocentrism, effected change in the world because they were so unlike the ideas already there. The method of science was important and promised to remake areas of life because it was not common sense. Scientists demanded attention because they knew things ordinary people did not, and perhaps could not.

The distinctness of science could also introduce tensions into attempts to forge connections with other parts of American culture. During the early nineteenth century, such connections were almost universally given, largely because the boundaries of science were so permeable. By the later 1800s, some authors stressed the limitation of science to the physical world and the in-ability of its devotees to speak on matters of more transcendent or immaterial importance. Yet it was still possible to bring science to bear on these subjects, and numerous Americans did, either by holding on to belief in the unity of all truth or by identifying truth with science and seeking to remake the world in a scientific image. Both of these options had remained viable during the late nineteenth century. But by World War I, they appeared somewhat less robust. Recent scientific truths such as relativity hardly seemed harmonious with any knowledge relevant to everyday life. One critic compared some of the implica-tions of relativity to telling a sports enthusiast that a new athlete could jump "fifteen thousand feet high."[145] Nor did it appear that any world inhabitable by humans could be wrung from its implications. Instead, attempts to make such scientific ideas or the methodology that gave rise to them intellectually, socially, or culturally relevant often smacked of illegitimacy, which threatened to breach the boundaries that separated science from other realms or to attract charges of indulging in the occult, mystical, or weird.

As the century progressed, the influence of science in the world came to rest increasingly on technology and medicine rather than on the ability of sci-entific ideas to offer enlightenment about the nature of reality. Frederick Lew-is Allen claimed that science occupied a position of prestige among ordinary Americans during the 1920s because of the "new machines and devices which they owed to the laboratory."[146] Though Allen's perceptions were not neces-sarily accurate, he at least thought mechanical application rather than some intellectual benefit explained the cultural authority of science. Application certainly became a strong element in the work of popularizers. Imagery of sci-ence as a font of new technology appeared in its most undiluted form at both the 1933 and 1939 world's fairs. Professor Henry Crew noted the many pos-sible definitions of science that existed in people's minds. But he announced that the science to be put on display at the 1933 exhibition was "the Science which has aided industry."[147] At both the fairs, visitors were able to see the latest and promised products of science, from long-distance telephoning to cigarette-smoking robots. But other sites resonated with the material power of science as well. First in New York and Chicago and then in additional cities around the nation, new museums of science and industry promised to bring a strongly technological version of science to American eyes, while Thomas Alva Edison, whose reputation only seemed to grow as the early twentieth century progressed, remained in the pantheon of great scientists of the age.[148]

While steeped in journalistic zing, pronouncements about the material wonders that would surely follow from science contained a kernel of plausibility in light of the technological improvements that had begun to exert an intimate and powerful influence over the lives of the majority of Americans, far more than during the nineteenth century. Technological visions of science coincided with the increased availability of consumer goods supposedly based on laboratory research or on scientific theory. Radio was perhaps the most notable of these. By 1928, over seven million sets had been sold. But it was not the only device being celebrated. A 1931 article in *Good Housekeeping* by Elsa Einstein, reflecting on her impressions of life in the United States with her famous husband, focused almost entirely on the presence of labor-saving technology in the home. It was during the 1920s that companies such as General Electric and Westinghouse began to market their products to individuals. Before this, they had directed their sales primarily to other manufacturers.[149] The highly technological visions of science at the two world's fairs were intimately related to corporate advertising and public relations.

In the midst of a sea of images of a technologically powerful science, many popular commentators did not want to lose sight of scientific ideas. The language of pure (or basic) and applied science, initially forged in the late nineteenth century, provided one possible rhetorical tool for peeling apart the core of knowledge about nature from its amalgamation with more practical, material concerns. The sentiment that "pure science is the spring that feeds all the applied sciences" circulated widely among scientists and popularizers. Numerous authors held up the radio as an example of "theoretical research" affecting "practical life."[150] Talk of inevitable practical application appeared even in discussions of relativity. Numerous commentators in magazines assured their readers that Einstein's ideas would someday surely find some practical outlet, including new energy sources, the transmutation of metals, or even the achievement of antigravity. A 1929 article in *Popular Mechanics* showed a drawing of a futuristic flying machine cruising over the Statue of Liberty along with the caption "Is This the Airship of the Future? A Motorless Craft Flying through Space Because Insulated against Gravity; Einstein's New Theory Opens This Field to Speculation."[151] Though valuable for justifying the importance of scientific ideas, the rush to emphasize their application easily led to a sense that there was no such thing as pure science, at least none worth talking about in public. Indeed, the term "applied science," against which pure research was so often cast, declined in English-language magazine article titles over the early part of the 1900s.[152] This decline suggests a return to the early-nineteenth-century view that all science was applicable, but in a much more narrowly defined way.

Against its technological power, the intellectual content of science sometimes became obscured. In his *ABC of Relativity* (1925), Bertrand Russell noted that "it is astonishing how little of the real world is revealed by physical science: our knowledge is limited, not only by conventional elements, but also by the selectiveness of our own perceptual apparatus."[153] Similarly, in his *Limitations of Science* (1934) and elsewhere, J.W.N. Sullivan claimed that science concerned a limited region of experience, though he quickly asserted that this limitation existed "not perhaps, as a necessary consequence, but certainly as a matter of historical fact."[154] Other authors were less careful in their claims. A reviewer of a book called *Riddles of Science* (1932) took great pleasure in informing "devotees of the 'religion of science' that there are riddles of science for which there is as yet no scientific answer."[155] Sometimes the utopian and the narrow appeared on the same page. A 1935 article in *Newsweek* pointed out chemical research, such as investigation of synthetic rubber and work with mice that promised an antidote to cancer in the near future. It also included a quotation from a Notre Dame chemist who cautioned readers that "not all the biochemists in the world can explain how the vegetable—with sunbeam, a few quarts of water, and some carbonic acid gas—is able to make a bushel of corn or a peck of potatoes."[156] Beginning in 1919 with his *Book of the Damned*—the damned being all those phenomena that orthodox science excluded or was blind to—former newspaperman and dabbler in the occult Charles Fort became one of the most insistent champions of the limited nature of science. Though Fort remained largely unknown in his day, he was admired by a number of American literati, including the novelist Theodore Dreiser. In the wake of many humanists' and intellectuals' concerns over the disruptive power of science in the wake of the Great Depression, many of these figures became involved in the founding of the Fortean Society in 1932 in New York.

During the second quarter of the twentieth century, the most common images of science in American popular literature depicted a field apart from nearly every other branch of human knowledge. This separation helped both to emphasize the power of science and to make it remote. Perhaps lay readers did become interested in relativity because it offered something beyond science or because Einstein seemed more than a scientist. As the century progressed, however, such a level of popular interest in any aspect of scientific knowledge for reasons other than its practical application became rare. The deepening rhetorical gap between science and the world of the average person helped to motivate popularizers throughout the rest of the century. But it also provided a ready resource for those Americans who did not want to pay much attention to the scientific. In an age of expanding information and entertainment opportunities, depictions of science as inaccessible could justify ignoring it in favor of more relevant sources of data or guidance about the world. Even more than during the late 1800s, freedom for scientific practitioners, sheltered behind the boundaries of science, also meant freedom *from* them and their ideas. After the Second World War, as the rhetorical borders of science grew less and less permeable, those inclined to invest science with value beyond its ability to produce new technology occupied the fringe areas of popular discourse, and sometimes sought to construct a more expansive, open, and alternative science all their own.

Talk about relativity and its creator also continued to develop after 1945. Einstein's work was sometimes portrayed as the foundation of atomic weapons, an association that leant his image an added note of pathos. During the 1950s, relativistic physics also played a role in cosmological debate between the big bang and steady state models of the universe. And in the 1960s, there was a surge of interest in Einstein's work among a number of different groups. Discussion of relativity became an element of the emerging disciplines of the philosophy of science and the history of science. Textbooks and teaching resources also appeared, including Laurence Carr's *Relativity for Engineers and Science Teachers* (1960). Finally, there was another wave of popular treatments of Einstein and his ideas, such as cosmologist Hermann Bondi's *Relativity and Common Sense* (1964) and Martin Gardner's *Relativity for the Million* (1967).

Renewed attention to relativity and the new physics more generally did generate some science talk. At the very least, Einstein remained a model of the ideal scientific practitioner even after his death in 1955. A survey of schoolchildren aimed at discovering shared images of the scientist in the late 1950s found classical Einsteinian attributes, including wild hair, still alive.[157] But such imagery was not necessarily linked to talk about relativity. Instead, as relativistic physics settled deeper into orthodoxy, it offered a less intense example of talk about the nature of science. In an environment in which science was assumed to be more or less recognizable on sight, there was little motivation to talk about it except in extreme and marginal cases. This was already becoming the case during the early 1900s, and was much more true during the latter half of the century. One of the most prolific sources of Cold War–era science talk was also one of the strangest, namely, growing interest in unidentified flying objects.

Chapter 4   UFOs

In the Shadow of Science

In 1969, PHYSICIST EDWARD Condon denounced the spread of what he called "Scientific Pornography." From 1966 to 1968, he had led a University of Colorado effort, on behalf of the United States Air Force, to study unidentified flying objects (UFOs). Over that period of time, Condon became a highly visible and outspoken crusader against not only flying saucers but all things "pseudo-scientific." In his view, publishers going after a profit often distorted science, printing, promoting, and selling "pseudo-science magazine articles and paper back books" by the tens of thousands and even millions. He was particularly concerned about impressionable youngsters. In a 1969 address to the American Physical Society, Condon threw down the gauntlet. When the minds of children are at risk, he declared, "I do not believe in freedom of the press or freedom of speech." Instead, he thought that "publishers who publish or teachers who teach any of the pseudo-sciences as established truth should, on being found guilty, be publicly horsewhipped, and forever banned from further activity in these usually honorable professions. Truth and children's minds are too precious for us to allow them to be abused by charlatans."[1]

By the end of the 1960s, as Condon was all too aware, UFOs had become a topic of widespread interest. In 1952, a technical writer for Douglas Aircraft Company complained that "sauceritis has spread like a prairie fire across the nation, inspiring hundreds of articles in magazines ranging from *Harper's* to *Look*, radio and television round-table discussions, and a syndicated comic strip."[2] Many public commentators followed Condon in claiming that the cure to this disease rested with science. Skeptics such as

Harvard astronomer Donald Menzel sought tirelessly to uncover the basic optical illusions or natural phenomena behind saucer sightings by the application of "plain scientific reasoning."[3] Others saw the rejection of science itself as the cause of sauceritis. A Pennsylvania State University physics professor saw in popular enthusiasm for UFOs hints that "society in general thinks science has gotten too big and many people like to think that science doesn't have all the answers."[4] But UFO advocates did not cede scientific ground so easily. During the mid-1960s, University of Arizona meteorologist James McDonald, one of a number of aspiring "ufologists," declared that "UFOs are the number 1 problem of world science."[5] He and others later criticized the Condon report as something less than the best example of that science. One of the most prominent ufologists, J. Allen Hynek, went so far as to suggest that UFOs "might call for a rearrangement of many of our established concepts of the physical world that will be far greater even than" the one formerly required by Einstein's relativity.[6]

Though they used it for dramatically different purposes, the science that partisans in the debate over UFOs fought for possessed far stronger and less permeable rhetorical boundaries than anything constructed with the rhetorical tools of the previous two centuries. As always, details regarding the nature of science varied, but after World War II these details did so within a framework that almost universally asserted its separateness and specialness. In the shadow of these boundaries, UFOs took on an air of transgression unheard of a century before, for Condon, at least, verging on the pornographic. Many early-nineteenth-century phrenologists had felt persecuted by critics, but they did not often depict themselves as clamoring to get into the scientific world. Simply by applying the methods of science, they assumed that they had already joined it. Late-twentieth-century advocates of ufology—both members of the scientific community and those beyond its borders—frequently reflected on their outsider status. Phrenologists sought legitimacy through public recognition. Those who claimed the title of ufologist typically sought it through recognition by the orthodox scientific community. It was agitation among enthusiasts for an official but nonmilitary inquiry that had prompted the Condon study in the first place.

But as the inside of science crystallized into a more definable place, so did its outside. In times past, some observers had drawn rhetorical boundaries in ways that constricted the scope and influence of scientific ideas, methods, and practices, though they almost never set science completely aside. However incomplete, scientific truths remained harmonious with other truths. By the early 1900s, the place of science in that larger network of truth was coming under stress, and attempts to assert links between the scientific and other realms of culture often appeared to contaminate either science or nonscience. Science itself sometimes seemed limited even in its study of the physical world. Likewise, depictions of scientific knowledge often divested it of much of the higher meaning it once had. It was increasingly possible to assert that science did not reveal universal truth or possess transcendent intellectual value. While some hopeful ufologists struggled to draw the boundaries of science in ways that included themselves and their work, many UFO enthusiasts embraced their status as outsiders and looked for intellectual satisfaction in an alternative to "orthodox" science. A few abandoned what seemed to them to be an ossified, inaccessible, and narrow-minded science altogether and turned instead to self-consciously nonscientific ways of viewing the world, natural and otherwise.

Late-twentieth-century Americans who discussed the often formidable boundaries of science did so with a body of science talk that was perfectly suited to the task. The vast majority of portrayals of science involved invocations of "scientists" as a special class of people. Technology also loomed large as the primary expression of scientific power and relevance, often at the expense of more intellectual and cultural linkages. Indeed, the circulation of other kinds of science-defining tools that had traditionally been useful in building bridges between the scientific and the nonscientific declined. Alternately, some new elements of science talk spread. The growing prominence of "pseudo-science" in magazine titles suggested the degree to which scientific boundaries had hardened and the ease and regularity with which they were violated.[7] With such rhetorical tools at hand, Americans had to work much harder than before to talk about science as an important part of their lives, a task that increasingly fell to particular subcultures, all of whom seemed a little strange in popular depictions. As they had over the previous century and a half, those boundaries surrounded a science that was both powerful and remote. And as the boundaries of science became more prominent and less easy to breach, a science set apart became a science easily set aside.

Flying Saucer Invasion

The era of the UFO began on June 24, 1947, when Kenneth Arnold, a Boise fire equipment salesman and experienced pilot, saw several unusual airborne objects while flying near Mount Rainier in Washington State. Although he described the "craft" as crescent-shaped, he compared their motion to discs skipping over surface of the water—an analogy that gave rise to the term "flying saucer." In the wake of Arnold's encounter a host of saucers, and a variety of other shapes, appeared as "people turned curious or uneasy eyes towards the skies."[8] Sightings by pilots other than Arnold, including the tragic case of

Captain Thomas Mantell, whose P-51 fighter had supposedly crashed during the pursuit of a UFO, and an encounter by an Eastern Airlines flight crew with a glowing, cylindrical craft in the night sky over Alabama, drew particular attention. Despite the efforts of debunkers, Donald Menzel most prominent among them, another "flap" broke out several years later, in the summer of 1952, when UFOs were detected both visually and on radar screens descending en masse onto Washington, DC. The sightings prompted the scrambling of Air Force jets and the longest military press conference since the end of World War II.[9] Subsequent waves of saucers arrived in 1957, 1966–67, 1972–73, and 1978, and even during off years Americans continued to see strange things in the sky. Future president Jimmy Carter reported sighting a UFO in 1969.[10] All together, in the three decades after 1947, fifteen million Americans claimed to see something strange in the sky.[11]

There were certainly precedents for reports of unusual aerial displays. A few saucer sightings even predated Arnold's. But between the decline of ancient and medieval interpretations of such events as supernatural signs, wonders, and portents during the seventeenth century and their incorporation into the UFO phenomenon in the twentieth they often appeared as isolated incidents and were quickly forgotten. Charles Fort combed eighteenth- and nineteenth-century periodicals for numerous discarded references to strange objects seen in the sky. During the fall of 1896 and the spring of 1897, there was even a nationwide rash of "airship" reports from thousands of people across the country claiming to have seen lights in the sky, cigar shapes—sometimes with wings, and spotlights from above. The airship sightings contained many elements that would reappear fifty years later. Some witnesses explained their experiences to newspaper audiences by appealing to the work of unknown terrestrial inventors who had finally solved the problem of creating a flying machine. On one occasion, however, an airship that supposedly crashed in Aurora, Texas, was identified as originating from Mars, whose famous canals were in the process of making their own headlines. Alternately, many astronomers appeared in print to debunk sightings as misidentifications of Venus or Arcturus. Despite such attention, however, the airship sightings faded quickly from view. Few commentators connected them with later talk of "foo fighters," small glowing lights seen by Allied pilots over Germany during the 1940s.[12]

Though some public commentators expected Arnold's encounter to quickly fade away, a columnist in a 1950 issue of Time noted, the "flying-saucer yarn was much too good to die young."[13] Postwar skies were certainly more crowded and interesting than before, partially because of the growing prominence of American airpower, as demonstrated by the war and the creation of the Air Force in 1947.[14] During the 1950s, enthusiasts such as Willy Ley urged excitement even higher, upward to "the newest, most exciting field of science—the conquest of space."[15] With new sources of experience came new kinds of categories. Flying saucers and the more technical-sounding "unidentified flying objects" that gained preference in serious-minded circles joined aircraft, rockets, and eventually satellites. Flying saucers soon appeared in a number of strategic rhetorical locations, from the *New York Times Index* and *Readers' Guide* to *Webster's Dictionary*.[16] A 1947 Gallup poll suggested that 90 percent of Americans had heard of them.[17] Some UFO enthusiasts projected the category backwards in time as well, retroactively making sense of a whole range of historical events. Robert Unger, a former B-52 pilot and later technical writer for Republic Aviation, claimed that, "judging from the similarity of reports since 1597," aliens had been observing the earth for some time.[18] One of the oldest historical cases mentioned involved the biblical prophet Ezekiel, whose vision of the throne of God in the sky included a reference to "wheels within wheels." In *Flying Saucers* (1950), Menzel explained Ezekiel's sighting as a perfect example of optical effects called mock suns and glories. For others, however, the long history and consistency of UFO sightings helped to show they were not a fad or symptom of Cold War jitters. One author went so far as to translate Ezekiel's description into a design for a spacecraft, whose landing he might have seen. Another claimed that no noise but a rocket engine could have matched the one described by the prophet.[19]

It was through the newly formalized category of the flying saucer/UFO that unexplained aerial phenomena became a subject of scientific attention, particularly under the auspices of the Air Force.[20] Formal interest began with Project Sign in 1948. While at first some Air Force officers took the possibility of extraterrestrial UFOs seriously, later investigators focused more on debunking sightings and assuring the public there was nothing to fear. Signaling this shift in emphasis, the Air Force reorganized Project Sign as Project Grudge in 1949. Since the project names were classified, the general public, unaware of any change, called both "Project Saucer." The Air Force reorganized its UFO study again in 1952, the year of the DC sightings, as Project Blue Book. The CIA also expressed some early interest in flying saucers.[21] All this formal attention suggests the degree to which the rhetorical establishment of American science paralleled its social establishment during the several decades after World War II. The contribution of science to national security, demonstrated in such wartime developments as radar, submarine detection, and the atomic bomb, had involved scientists in what President Dwight D. Eisenhower ominously called the "military-industrial complex"—to which Adlai Stevenson later appended "academic." Throughout the 1950s and especially 1960s, researchers obtained training in an educational system increasingly supported by federal funds, ran or advised the agencies that distributed those funds, and worked in academic and industrial environments rich with government contracts. Edward Condon's career was common to many physicists of his generation. After working on the Manhattan Project, he moved on to become director of the National Bureau of Standards. His later experience with the Air Force, which provided over $500,000 to the University of Colorado, was likewise typical of the system of federally supported research.[22]

The prominence of the established scientific community in debate over the UFO phenomenon grew to new heights in the 1960s. While media accounts of the Air Force investigation made frequent reference to unnamed "scientists," Projects Sign, Grudge, and Blue Book used only a skeletal contingent of actual researchers, including J. Allen Hynek, who had not yet begun to take UFOs seriously. At its height, in fact, Blue Book consisted of two officers and a sergeant in a single room at Wright-Patterson Air Force Base. By the late 1960s, Hynek had emerged as a strong critic of the Air Force program, its attitude, and its methodology.[23] Though often considered the founder of ufology, Hynek was not the first scientist to express a positive interest in UFOs. Fellow astronomer Jacques Vallee, for a short time Hynek's colleague at Northwestern University, had begun to write about UFOs during the mid-1960s and early on mentioned an "invisible college" of scientists who were fascinated by the phenomenon but not yet ready to identify themselves.[24] Meanwhile, a number of skeptical fellow travelers, most notably the astronomer Carl Sagan, joined Menzel. Such involvement paralleled increasing frustration with military control of UFO research among civilian groups, which had begun to press for congressional hearings on the subject. A major series of sightings in Michigan in 1966 helped to increase concern over the Air Force's investigation. The suggestion offered by Hynek, who was still affiliated with Blue Book, that phosphorescent swamp gas might explain what witnesses saw, received nearly as much ridicule as flying saucers themselves. Subsequently, Gerald Ford, then a Republican representative from Michigan, added his voice to calls for congressional attention.[25]

By the late 1960s, with criticism on the rise, the Air Force had become eager to get out of the UFO business, and sought several major universities to conduct a definitive study of the phenomenon. After a number of refusals, the Air Force finally secured the cooperation of the University of Colorado. After fifteen months, the massive Condon report concluded that, despite only partial success in offering definite explanations for the sightings considered by the study, there was no evidence of any extraterrestrial visitation to earth, that the study of UFOs had contributed nothing to the advancement of science, and that the Air Force should not be preoccupied with chasing flying saucers.

Rather than quelling debate, the project's findings met with a barrage of questioning. Even before the study had been completed, detractors had emerged both within and without the study's ranks. Articles critical of Condon and the work he directed appeared in *Look* magazine and in *Science*.[26] Nor did the Condon report signal the end of established scientific interest in UFOs. A survey of the 1,356 members of the American Astronomical Society during the mid-1970s revealed that 53 percent believed that UFOs deserved scientific study, while sixty-two astronomers claimed to have witnessed one.[27] However, most active work on the UFO phenomenon increasingly occurred outside of the structures of orthodox science.

Word of the Condon report and of scientific debunking of UFOs more generally often made its way into the public arena through a highly organized industry of science popularization, one that mirrored the formalization of scientific practice. Indeed, the two were related. Though the barrier of secrecy prevented some information from reaching the public, reverberations from the role of science in World War II and its perceived importance in the Cold War helped to boost popularization. Just as journalists and scientists had joined forces in the 1920s to increase the visibility of science, an alliance of publishers, scientists, professional societies, and government agencies made a push during the 1950s to improve the "public understanding of science," a phrase almost universally synonymous with popular appreciation of science. Newspapers and magazines employed reporters and editors dedicated to science news. Scientific information also appeared on the newest medium, television. Aside from news segments that covered such big stories as the moon landing in 1969, there were also special programs focused on science. During the early 1950s, for instance, CBS aired a Sunday morning science show called *The Search*.[28] Popularization continued to flourish in a variety of media in subsequent decades. The science series *Nova* began in 1973. And in the early 1980s, another "boom" in popular science occurred.[29]

Observers have sometimes linked the darkening of American skies with flying saucers, particularly during the 1950s, with a postwar "scientific craze."[30] Certainly, many information consumers valued, enjoyed, and sought out scientific information. But, just as earlier in the century and despite the efforts of popularizers and their frequent claims about widespread demand for scientific information, science often occupied a very modest profile in the larger context of postwar news and entertainment. Though the number of magazine articles about science listed in *Readers' Guide* did rise a little over the 1900s, they never amounted to more than 1 percent of the total number of articles listed per year.[31] Attempts at popularization garnered lasting success only when they curbed their desire for wide audiences. Among a number of new science magazines begun during the 1950s, only the revamped *Scientific American* survived, in large measure because it appealed to a fairly limited technical readership.[32] A 1957 survey sponsored by the National Association of Science Writers confirmed the low profile of science news in the mass media, and the development of a specialized audience, most broadly those with high levels of educations, for scientific information.[33] The magazines begun in the popular science boom of the 1980s also survived insofar as they appealed to limited audiences.[34]

Meanwhile, a great deal of the popular discussion of UFOs occurred outside and sometimes even in opposition to the established system of popularization. UFOs found a place in the more or less orthodox programs of radio commentators such as Frank Edwards and Walter Winchell and in a 1950 CBS documentary hosted by Edward R. Murrow. But they also appeared in less typical sources of science news as well. A 1950 article by retired Marine major Donald Keyhoe in the men's magazine *True* was, according to one historian, "one of the most widely read and discussed articles in publishing history."[35] But talk about flying saucers continually threatened to spill out of control. Many other believers and skeptics created their own "galaxy of books" and articles as sightings multiplied. Enthusiasts and organizations sometimes turned to or founded independent publishing houses and periodicals, such as *Saucer News* and the *UFO Newsletter*, dedicated to revealing information that the orthodox press ignored. Llewellyn Publishing, a firm originally founded in 1901 and focused on astrological literature, broadened its coverage with the inauguration of *Fate Magazine* in 1948. The first issue featured an article by Kenneth Arnold reflecting on his airborne encounter.[36] Subsequent issues delved more deeply into the UFO phenomenon. The roots of UFOs in unorthodox literature ran deep. Flying saucers first emerged as a concept from the pulp magazines of the 1930s and 1940s, particularly in Ray Palmer's *Amazing Stories*. Palmer later helped to co-found *Fate*.[37]

Ordinary citizens also created their own institutional structure from which to respond to and with which to legitimate the UFO phenomenon. The National Investigations Committee on Aerial Phenomena (NICAP), founded in 1956, occupied the most prominent position among independent UFO-focused groups during the 1950s and 1960s. Keyhoe, its inspiration and leader, represented perhaps the most influential investigator, aside from Menzel, in the decade after World War II. Whereas Menzel often explained modern saucers in terms of optical illusions, such as temperature inversion in the atmosphere that could bend light and radar signals, Keyhoe's military contacts led him to believe that saucers came from outer space, where aliens had stepped up their observation of our planet in the wake of the first atomic bomb. He also claimed that the military knew perfectly well the origin of UFOs, but that they were actively covering up that conclusion to prevent mass panic. Throughout the next several decades, NICAP tirelessly promoted UFOs as a serious national security and scientific problem, a role that often placed it in an uneasy and frequently tense relationship with the Air Force.[38] NICAP was not alone, however. After seeing a UFO in New Mexico and later reading about Kenneth Arnold's encounter, Coral Lorenzen, a young house-wife from Sturgeon Bay, Wisconsin, decided to found the Aerial Phenomena Research Organization (APRO). Similarly, a Bronx housewife named Marylin Shaw founded an organization called Civilian Saucer Intelligence after she saw something strange in the sky.[39]

Perhaps the most interesting and unorthodox aspect of the UFO phenom-enon during the 1950s involved the "contactee" or "saucerian" movement. Centered largely in Southern California, the contactees amalgamated ele-ments of popular culture, heterodox theology, and science into claims of very close encounters with the flying saucers and their occupants. George Adamski, who had been deeply interested in Eastern religions, Theosophy, and psychic phenomena before turning toward the skies, became one of the most promi-nent contactees. In several books during the 1950s, which included close-up photographs of flying saucers, he described conversations with technologically and spiritually advanced aliens from Venus, Mars, and Saturn. Like many of the other aliens with whom contactees talked, Adamski's handsome, blonde-haired "space brothers" expressed concern about the fate of a human race that now had the atomic bomb at its disposal. In a survey of American UFO groups conducted in 1967, Jacques Vallee found a fair amount of skepticism regard-ing the contactees' claims, but many people did believe them, and the issue of interaction with UFO occupants often divided researchers and organiza-tions who wanted to appear legitimate from those that had little interest in becoming mainstream. The Amalgamated Saucer Clubs of America offered a contactee-friendly environment, while NICAP refused to consider occupant reports from any source, preferring accounts of more aloof visitors from afar.[40]

Popular attention to UFOs also represented a much larger postwar surge in discussion of unorthodox ideas that comprised Condon's scientific pornog-raphy. *Worlds in Collision* (1950), by Russian émigré Immanuel Velikovsky, was one of the first works about such topics to gather widespread attention. *Worlds* offered explanations for the seemingly miraculous disasters of the Bible in terms of cataclysmic astronomical events that surpassed anything that contemporary astronomers were prepared to accept.[41] Other icons of the unorthodox came to public attention in subsequent years. Nature writer and researcher Ivan T. Sanderson began collecting sightings of what would even-tually be called "Bigfoot" during the 1950s. His *Abominable Snowman: Legend Come to Life* appeared in 1961. Sanderson later turned to the investigation of

UFOs. Shortly afterwards, journalist Vincent Gaddis introduced readers to the mysteries of the Bermuda Triangle in his *Invisible Horizon* (1965). By the end of the 1960s, there was an explosion of interest in strange phenomena, particularly on college campuses. In the introduction to papers presented at a 1969 symposium on UFOs sponsored by the American Association for the Advancement of Science, Carl Sagan united concern over what he perceived to be a "drift away from science" among the younger generation with "a range of borderline subjects that have high popularity among these same people—including UFOs, astrology, and the writings of Velikovsky."[42] Astronomer Thornton Page similarly connected interest in UFO reports among college students with their desire to "'have a cause,' opposing US government policy, or university faculty, or 'The Establishment.'"[43] Interest in the unusual also spread beyond college campuses. The *National Enquirer*, which had gained substantial success during the 1950s through sensational crime stories and Hollywood gossip, turned toward unusual and supernatural events in 1968.[44]

The growing prominence of unorthodox topics in American public discussion during the late 1960s in turn coincided with the expansion of a wide variety of "countercultural" movements.[45] Many of these movements shared a rejection of what their members perceived as the too restrictive social and cultural norms of postwar America and sought to inscribe new religious, political, and sexual boundaries, often through the pages of an alternative "underground press." Science appeared in the sights of Americans interested in questioning the status quo, as well. The specter of environmental damage raised in Rachel Carson's *Silent Spring* joined the potential of nuclear annihilation in prompting a more critical view of the impact of scientific knowledge and applications on the world.[46] Many left-leaning humanists and politicians attacked the scientific establishment as having come to represent the elitist interests that sought to control that nation rather than the detached quest for the truth about nature. Observers inside and outside the scientific community expressed concerns about the influence of military money, concerns sharpened by debate over the war in Vietnam. Cultural historian Leo Marx claimed in 1970 that "the scientific and technological professions have in fact strengthened the ideology of American corporate capitalism, including its large armaments sector."[47] Meanwhile, many members of the counterculture expressed interest in the transcendent possibilities of astrology and other unorthodox sciences.

Mixed feelings about established science spread into the general culture as well. Beginning in the mid-1960s, an unprecedented note of criticism entered public discussion of science. This critical tone was part of a more general trend among journalists to take an aggressive and questioning stance, though science writers did so less than other sorts of reporters.[48] Public opinion also seemed more critical of science. Between 1966 and 1973, according to one survey, confidence in science slipped 19 percentage points, compared to a loss of only 6 points for religion. Again, this reduced confidence was part of a larger move toward pessimism regarding fundamental American institutions.[49] Eventually, established science came under institutional attack. Physics in particular suffered a degree of disestablishment during the early 1970s. Federal funding for research and development declined 18 percent in real dollar value between 1967 and 1973. Pentagon support for basic science declined in absolute dollar value over the same period.[50] The system of science advisors that had been constructed in the 1960s fell victim to the bitter politics associated with Vietnam and the Nixon administration. The scientific establishment recovered somewhat during the 1980s, as the Cold War flared. But its fortunes fell again during the early 1990s, in the wake of the fall of the Soviet Union and widespread reevaluation of federal and military spending on research and development.

Despite the negative conclusion of the Condon report, the UFO phenomenon also continued to develop during the 1960s and 70s. When the Condon report cast a shadow over attempts to integrate work on UFOs into mainstream governmental and academic institutions, 350 delegates from organizations such as NICAP, APRO, and the Mutual UFO Network (MUFON) met in 1975 at a conference in Fort Smith, Arkansas, and agreed to pool their findings in the J. Allen Hynek Center for UFO Studies (CUFOS).[51] Skeptics, who eagerly talked about topics most mainstream researchers hoped to ignore, also organized themselves. In 1976, a collection of well-known doubters led by philosopher Paul Kurtz formed the Committee for the Scientific Investigation of Claims of the Paranormal (CSICOP) and eventually began publication of its associated journal, the *Skeptical Enquirer*. Americans' experience with strange phenomena also developed in new ways. "Close encounters," often involving electrical disturbances to car engines and radios or physical effects such as burns, became increasingly common and strange. In therapy sessions involving hypnosis, Betty and Barney Hill recalled being abducted by short aliens with large black eyes while driving in New England in 1963. These were not the friendly, long-haired, spiritually advanced beings with imperative messages to send to humankind, as described by the contactees. Both of the Hills depicted a frightening and bewildering experience, complete with what seemed like a medical examination, lost memories, and missing time. Many subsequent abductees also reported paranormal experiences, such as the supposed development of psychic powers. These accounts blurred the lines between "hard" encounters and those that many serious ufologists and skeptics alike dismissed as silly.[52]

The Age of the Scientist

Debate over UFOs often centered on the authority of recognized scientists, such as Condon or Hynek. But popular depictions of scientists demonstrated how the trends of the previous century and a half of boundary-making both provided a source of authority and helped to limit it. During the 1800s, those limits had largely corresponded with the borders of the physical world. Meanwhile, discussion of men of science invested them with moral attributes and a general character that helped to ensure that, if done correctly, their work would harmonize with a larger realm of truth. Over the next several decades, science sometimes seemed confined to particular pieces of physical nature remote from the world of ordinary experience. Scientists themselves became increasingly distinguished from other sorts of people, often through impersonal qualities, such as their use of a special method or pieces of specialized equipment. Einstein represented in some ways a throwback to an earlier model of the scientific practitioner, and, as time wore on, his ability to straddle the divide between science and the rest of culture became a rare gift. Burnham recounted the sad declaration of one late-twentieth-century observer that "Einstein is dead," meaning that the "broad, culturally responsible" scientist had passed away.[53]

Nevertheless, "scientists" became one of the most important keywords of scientific power and legitimacy in popular science talk over the 1900s. In 1962, chemist Sydney Ross claimed that in the modern era "an abstraction named 'the scientist' has been given form in people's minds as a new figure of authority, corresponding to the priest or witch-doctor of a more primitive culture, whose 'scientific' statements can be accepted with child-like faith."[54] Whether this was actually true or not, Ross believed it, and so did others. In mass media coverage, scientists stood in as the public face of science and the source of scientific statements. The late-nineteenth-century assertion that "science says" commonly gave way to "scientists say" and invocations of scientists occupied an increasingly prominent role in magazine article titles through the 1950s.[55] Even when widespread deference to the "assurances of 'experts'" such as Menzel was criticized, it was not generally abandoned. Author and UFO researcher Don Berliner complained that in a 1978 article in *Science Digest*, "if a scientist says light rays can be bent to make headlights look like glowing objects overhead, we confidently accept that some mechanism must exist to make it so." But rather than turning his back on expert testimony, Berliner invoked his own authorities, noting a recent report from the Environmental Technical Applications Center of the Foreign Technology Division of the Air Force that called into question the ability of Menzel's temperature inversion theory to account quantitatively for visual sightings of UFOs.[56]

Rhetorically at least, observers on all sides of the UFO phenomenon treated those included under the label "scientist" as members of a special, and typically closed, group. Few outside the boundaries of the scientific community had depicted themselves as scientists amidst debate over relativity, perhaps reflecting the technical difficulty of the theory. But such restraint was also evident in public discussion of UFOs, in which public commentators routinely challenged the ideas of scientific practitioners. Contactee George Adamski called himself a "scientist." By contrast, he charged Menzel with being "scientifically dishonest" and of hiding the truth behind a "mass of astrophysical jargon that sounds convincing to the uninitiated, merely because it happens to be 'scientific.'" Those who engaged in that behavior, he suggested, "should drop all pretensions to the title of 'scientist' which implies 'one who knows' and 'one who thinks.'" The true scientist was also a philosopher.[57] But Adamski was one of the only people interested in UFOs to claim to be a scientist without advanced training in some branch of science. Broad and weakly bounded definitions such as his became all but impossible to maintain after mid-century. A similar reluctance to use the title of scientist appeared among other unorthodox groups, too. The antievolutionist Presbyterian minister Harry Rimmer might have declared himself a "research scientist" in the 1920s, but later devotees of creation science relied more on formal education to make the bearers of their ideas scientific.[58]

Surveys conducted between the late 1950s and 1970s further suggest that, for many Americans, scientists composed a group separated from ordinary society, often characterized by its narrow focus on science to the exclusion of other parts of life. Such perceptions were not necessarily negative. In a 1958 survey conducted for the NASW by the Survey Research Center of the University of Michigan, scientists cut a very positive image. Their substantial intelligence, followed by references to their high levels of education, led the list of defining characteristics offered by respondents. A 1957 survey of American high school students designed by Margaret Mead and Rhoda Metraux similarly found scientists described as geniuses that had benefited from years of training. Few of those queried in the NASW poll offered any negative impressions, but 41 percent did believe that scientists were likely to be "odd and peculiar people."[59] The Mead-Metraux survey revealed a far greater sense of oddity. The images of scientists they discovered often appeared devoid of connection to other people, including their families, or to anything outside of science. Some students suggested that scientists, given their science-obsessed existence, should not even marry. Later polls portrayed scientific practitioners as working "in isolated laboratories." A 1975 survey found that while those who identified themselves as scientists often saw themselves and their fellows as "typically approachable, sociable, open, unconventional, socially responsible, and popular with broad interests," nonscientists generally viewed them as the opposite.[60]

Scientists' focus on science to the exclusion of all else could justify their place as experts, and their detachment from society could reinforce a view of them as paragons "of reason and objectivity."[61] Their reliability often stood in stark opposition to that of the general public, even among UFO advocates. In 1973, Vallee lamented that it was difficult to explain science to the average person under the best of circumstances, and much more so in a new field such as UFO research amidst controversy and "widespread public misunderstanding."[62] The closing of orthodox scientific circles in the wake of the Condon report may have further prompted those who wanted to be seen as experts on UFOs to join skeptics in using imagery that stressed the unreliability of the public and their susceptibility to suggestive influences. The notion of an impressionable public, in need of constant protection and mindlessly responsive to reassurance and instruction, also helped to justify the established political, military, commercial, and scientific authorities of the Cold War United States more generally. During the mid-1950s, for instance, media analysts had "began to argue that the United States had become a 'mass society,' characterized by atomization and the isolation of individuals. The declining importance of primary groups, they believed, enhanced the direct influence of the press."[63]

Among debunkers, popular unreliability tended to reach its zenith, or nadir, in discussion of UFO witnesses, whom they often portrayed as marginal, even mentally unstable, figures. A 1952 *Time* article claimed that "many of the 'sighters' are newspaper delivery boys, excitable old ladies, and other people with no technical training." Eleven years later, in almost identical language, a *Newsweek* article attributed some of the strangest sightings to "excitable old ladies and imaginative schoolboys."[64] Skeptics attributed widespread belief in the reality of UFOs to the enormous "national capacity for gullibility."[65] Numerous media articles critical or dismissive of the flying-saucer craze explained sightings by portraying it as the malady of a populace whipped up by talk of space travel, lurid science fiction, general atomic-age anxieties, and hoaxes. Skeptics often emphasized the latter in their explanations of flying saucers.[66] By the same token, many scientists and those sympathetic to their expertise frequently argued that scientifically trained observers made the trustworthiest witnesses of things in the sky. In magazines such as the *Bulletin of the Atomic Scientists* or *Science News Letter*, Sagan and other debunkers claimed that "the most reliable testimony is that of the professional astronomer."[67] Likewise, those scientific insiders who followed Hynek in hoping to bring UFOs within the fold of established science stressed the need to get reports from "trained scientific observers" and happily accepted astronomers' testimony that they had seen things they could not immediately explain.[68]

But a separate and narrower image of scientists provided grounds for dismissing their assertions as well. A 1967 *Time* article suggested that "many laymen, baffled by the scientists already, might find the overthrow of all their lore," in this case the impossibility of exceeding the speed of light, "quite entertaining."[69] Some UFO enthusiasts called attention to the supposed demonstration by late-nineteenth-century astronomer Simon Newcomb of the impossibility of heavier-than-air flight to show the potential narrow-mindedness of the orthodox scientific community. The oldest and perhaps most widespread such tale involved the "notorious boo-boo of the French academy in 1790," when it rejected the notion of meteorites in the face of testimony by ordinary farmers that they had seen stones fall from the sky.[70] Hynek, as a member of the scientific community as defined by training and occupation and a critic of its general refusal to consider the problem of UFOs, frequently used rhetorical tools that both heightened the prestige of scientists and limited it. In some of his public comments Hynek stressed the need for "professional" attention to the flying saucers.[71] At other times, he was willing to sever almost entirely his colleagues' possession of the title of scientist from their right to speak for science. In a chapter of *The UFO Experience* (1972) critical of the Condon report, he warned his readers that "science is not always what scientists do."[72] Curtis Fuller, editor of *Fate*, echoed Hynek in wishing that "scientists were more scientific."[73]

Scientists were perhaps, now and then, less than scientific—whatever that meant—but they retained a general ownership of science. In discussion of the UFO phenomenon, scientists' main rivals in the making of truth statements about the world were often witnesses. The willingness to cast witnesses and scientists into opposing roles sometimes took on a populist, even defiant, tone. A 1958 *Time* article noted the UFO encounters of "preachers, military personnel, engineers and just plain folks."[74] In his *Interrupted Journey* (1966) and *Incident at Exeter* (1966), journalist John Fuller focused on telling the stories of "articulate and reliable citizens."[75] Many books about UFOs aimed at popular audiences did little more than provide an accumulation of stories about sightings as experienced by more or less average individuals. The 1966 sightings in Michigan provided one of the most notable examples of ordinary folk versus the dismissive and haughty scientific community. The explanation offered by the Air Force through Hynek, that glowing marsh gas had been responsible, "satisfied none of the people who had actually seen the object." Frank Mannor, a Michigan farmer who made the initial report of something strange in the sky, asserted in *Life* magazine that "I'm just a simple fellow. But I seen what I seen and nobody's going to tell me different."[76] Yet, even if they were right, witnesses were never called scientists, nor did supporters label their observations as science. Science required more than mere truthfulness.

The Scientific Establishment

In discussions of UFOs, the right to membership in the scientific community had little to do with a scientific spirit or a lust for truth. Instead, one of the most powerful rhetorical tools for delimiting boundaries around that community involved highlighting the institutional affiliations of scientists. Association with formal groups, such as the federal government and universities, had become a commonly used tactic to establish scientific legitimacy before 1947, and it received a significant boost in the Cold War era. The incorporation of scientific workers into powerful national institutions lent a much greater definiteness to the scientific community as a whole, even as authors of popular literature often described its members in faceless terms. Such definiteness appeared most intensely in occasional references to the "scientific Establishment" or the "government-military-scientific 'establishment,'" terms that seemed to suggest a far less open set of boundaries than earlier depictions of a "scientific world."[77] This was particularly true during the late 1960s and 1970s, when "establishment" was often a loaded word, suggestive of anything but inclusiveness. The strong presence of institutional links also narrowed the ground on which scientists stood and the places in which science was depicted as taking place. To many commentators who wanted more access to scientific knowledge, the links between scientists and official institutions seemed to separate science even further from the world of ordinary experience, not behind an increasingly technical and mathematical form, but beneath a shroud of official secrecy.

The importance of institutions was clearly visible amidst debate over UFOs. Among those people cited in late-twentieth-century magazine articles about UFOs as sources of facts or explanations, slightly over 50 percent were linked to an institution, either through occupation or education. The number was 5 percent higher among sources explicitly identified as scientists or with disciplinary labels, such as "physicist."[78] Americans left unprotected by institutional affiliations often sought to find shelter in organizations of their own, such as Donald Keyhoe's NICAP, APRO, or MUFON. Meanwhile, focusing on institutional affiliation corresponded with a lack of attention to scientists as individuals. Those doing science themselves often became anonymous figures, almost entirely devoid of the character that had once mattered so much. Even during the early 1900s, mass media depictions of scientific practitioners had begun to portray them as faceless members of organized teams or research groups.[79] An appeal to the personal characteristics of scientists generally occurred only when added emphasis was required. A *Saturday Review* article strongly supportive of Condon praised his morality, passion, and humor, but not until establishing his connection with such institutions as the Bureau of Standards, the American Physical Society, the National Academy of Sciences, and the University of Colorado.[80] Scientists often became anonymous in literature, too. A recent review of twentieth-century science fiction noted the relative absence of scientists as characters after the late 1950s. As time progressed, they became a "faceless force, as anonymous as scientific-technological progress itself."[81]

Among established institutions, the military provided the most common reference point in the identification of reliable sources of information during the 1950s. The Cold War equation between national security and science helped to encourage this linkage. Journalists were often quick to portray Projects "Saucer" and Blue Book as bringing the "finest technical equipment" to bear.[82] Observers also made frequent mention of "Air Force scientists" or of the combined judgment of "scientists and Air Force officials." Correspondents claimed that "a selected scientific group under the supervision of the Air Force" or "a panel of top scientists" at work for the military was studying reported sightings.[83] The tendency to locate the most advanced scientific equipment and expertise in the armed forces boosted the military's perceived abilities, but also made scientific claims appear more serious. The term "UFO," so enthusiastically accepted by those seeking to avoid the derogatory "flying saucer," originated in military jargon. Acronyms, including NICAP, APRO, and MUFON, also invoked a military aura to achieve some measure of orthodoxy.

Whatever the advantages of a strong association with the scientific establishment, close links between science and the federal government, particularly the military, raised concerns among many Americans, including scientists.[84] In the years leading up to the Condon study, a wide variety of people, from senators to journalists, called for congressional hearings to examine the topic. Others suggested that the investigation be taken over by NASA or that the federal government appoint a panel of experts to analyze the most puzzling cases. After the Air Force announced the closure of Project Blue Book in

1969, Stuart Nixon, then spokesperson for NICAP, expressed relief that UFOs could finally "be given the serious scientific attention they require, free from military considerations."[85] But those seeking to bring UFOs within the fold of established science did not return to older equations of doing science with internal attributes, or abandon reliance on institutions. Numerous ufologists continued to hope for some involvement from the scientific establishment in further research. In 1981, Hynek noted that "we would never have gotten to the moon if we had merely given the project to volunteers to work on during the weekends."[86] Skeptics, such as Condon, safely ensconced in the institutional network of Cold War science, ridiculed ufologists' desire for federal money or a "National Magical Agency."[87] Of course, the federal government more generally suffered from growing levels of suspicion during the late 1960s, prompting other public commentators to stress the reliability of academic institutions and private organizations, such as NICAP or CUFOS. When the Air Force sought to find some means of escaping growing charges of a cover-up and to justify the closure of Project Blue Book, it turned to universities. Once the research contract was announced in 1966, skeptics and believers alike expressed hope that finally the truth might come out. Astronomer and debunker Philip Wylie happily announced, "[N]ow that the University of Colorado has been engaged to study flying saucers, it is possible to think of them in a new way: scientifically."[88]

The reliance on institutions to bestow a measure of trustworthiness, even if only because they provided the money and resources to get to the truth, represented a much larger trend in the discussion of Cold War sources of information.[89] Such associations relieved the average information consumer of having to invest time and effort in judging claims with reference to less tangible factors, such as methodology or character, which were often inaccessible and missing from mass media accounts in any case. Institutions even played a role in authors' judging the testimony of UFO witnesses. Among authors who drew attention to the sightings of ordinary people, encounters by pilots and police officers held particular value. In one of his first UFO articles, John Fuller invested a series of sightings in 1966 near Exeter, New Hampshire, with particular importance because they had been reported by two police officers "whose character and reliability seemed to be unassailable."[90] Representatives of law enforcement stood as "credible witnesses on everyone's list."[91] Similarly, Senator Barry Goldwater declared in 1979 that "I've never seen a UFO, but when Air Force pilots, Navy pilots, and airline pilots tell me they see something come up on their wing that wasn't an airplane, I have to believe them."[92]

But an easy means of judging claims, including those of scientists, could also become a reason for suspicion. Ted Bloecher, director of the research division of Civilian Saucer Intelligence, an organization of UFO enthusiasts based in New York, complained in 1955 about the "hypnotic effect" of the words "Harvard University" behind the name of skeptic Donald Menzel. This magical incantation had "blinded more than one reviewer," Bloecher charged.[93] Others suggested that the public was being blinded in more sinister ways. Claims of conspiracy and cover-up attended popular discussions of UFOs almost from the beginning, though they became far more pervasive amidst the eruptions of antiestablishment feeling during the late 1960s and early 1970s. Radio commentator Frank Scully's *Behind the Flying Saucers* (1950) anticipated the same basic storyline as the tales about Roswell, New Mexico. Relying on information from a supposed government scientist identified only as "Dr. Gee," Scully asserted that not only did the Air Force know much more about flying saucers than it let on, but that the military actually had recovered three crashed saucers and was busily, and secretly, working to understand their technology. Five year later, Donald Keyhoe published his *Flying Saucer Conspiracy* (1955), in which he laid out his argument that Air Force scientists and officials were actually aware of the true extraterrestrial nature of the UFOs, though without having actually recovered any alien technology. As many an author would later echo, the Air Force chose not to announce anything for fear of mass panic. Such paternalistic concern could easily shade into more sinister motivations. Scully warned his readers that "anything remotely scientific has become by government definition a matter of military security first; hence of secrecy, something which does not breed security but fear."[94]

Concerns over secrecy sometimes arose in broader discussion of science and scientists. As members of a special group, endowed with powerful knowledge but without the same sorts of social connections that helped to inhibit and control the behavior of ordinary people, scientific practitioners became implicated in many sorts of conspiracies. A review of postwar surveys of scientists' depictions suggested that young children often imagined them as working on "secret and destructive" projects.[95] Amidst an atmosphere charged with worry over Communism, concern over what scientists were really doing and thinking could take on a serious political dimension as well, as with the fall of Robert Oppenheimer. Edward Condon himself became the target of the House Un-American Activities Committee in the late 1940s because of his liberal political views and somewhat abrasive character. Even when scientists were not tainted by Communism, they could become the focus of other suspicions that they concealed something. When Hynek, then still the scientific voice of the Air Force, was asked if the 1966 Michigan sightings involved a secret aircraft, a columnist in *Life* magazine reported his reply in almost conspiratorial tones: "'I think I know more of what is going on than . . . ,' he began, but then halted and said, 'so I don't think I should say anything. . . . I'm sure there is some natural explanation for all of this.'"[96]

Science and Technology

Talk about the secrecy that sometimes shrouded the most advanced science was also a testament to its perceived power. Cutting-edge technology in the service of national security became a staple of spy movies, where white-suited technicians provided protagonists with watches carrying laser beams and cars that could transform into submarines. Yet just as many images of those who did science had grown narrower, so too did ideas about the nature of scientific power in the world. During the 1830s and 1840s, the utility of scientific knowledge had seemed to know little limit. Over the next century, a strong note of mechanical application became increasingly audible in science talk, though even by the early 1900s it did not yet completely dominate discussion of science. Widespread attention to the intellectual value of scientific knowledge in realms from the ethical to the social continued. In the decades after World War II, the broad intellectual relevance of some parts of science became difficult to discuss, and many popular depictions of science stressed its presumed material benefits alone. That tendency was sometimes attributed to the spectacular display of scientific might over Hiroshima and Nagasaki. A 1952 article in *Popular Science* noted that "our imaginations have been working overtime ever since the A-bomb showed us what science can do."[97]

In the light of the mushroom cloud, the technological power of science was frequently depicted as an inevitable force and a symbol of progress. Astronomer William T. Dearborn, a colleague of Hynek, noted that extraterrestrials, "just a cosmic clock-tick ahead of us in achievement, would have not only inconceivably advanced scientific ability but technological skill beyond our comprehension."[98] The presumed high technology of aliens overlapped and bled into visions of a human destiny awash in technological wonders. A doubter succinctly lamented in 1959 that believers in extraterrestrial UFOs had "appropriated the world of the future."[99] That charge was not far off the mark. An advertisement for the car-care product Gumout in a 1997 issue of *Popular Mechanics* included a sequence of automobiles from the 1960s onward. The image for 2032 was a flying saucer. More seriously, some UFO enthusiasts, following the lead of Jacques Vallee, suggested that "UFOs are teaching devices leading mankind into the future," perhaps by demonstrating the advanced technology we may one day use ourselves.[100]

Whether UFOs were in fact specimens of advanced science and technology, they did serve as vehicles for an expanded technological vocabulary. One journalist's comparison of international reports of UFOs indicated that the focus on technological aspects of sightings seemed a particularly North American response. Reports of encounters from the United States in 1954 dwelt on descriptions of aerial vehicles; prominent French sightings revolved around contact with their occupants.[101] The expanding genre of science fiction also helped generate a growing vocabulary with which Americans could make sense of the physical power and potential of scientific knowledge. Eventually, a semiformal body of knowledge about powerful alien technologies emerged at the intersection of science fiction and nonfiction UFO books such as Scully's *Behind the Flying Saucers*. In a reflection of the advanced technical knowledge of professional engineers and scientists, those discussing alien technology wielded their own specialized terminology, from motherships to antigravity generators to magnetic propulsion. Nor were such speculations limited to the UFO phenomenon. They became ubiquitous in science fiction, where such ongoing sagas as *Star Trek* and *Star Wars* generated a host of technical manuals on the arcana of warp drives and light sabers. Elements of this futuristic technological vocabulary often had remarkable stability, crossing the boundaries between different realms of speculation. A device called the "Beefield-Brown generator," which was often explained with reference to electromagnetism, the work of Nikola Tesla, and Einstein's unified field theory, appeared in discussion of UFO propulsion, antigravity, and the so-called Philadelphia Experiment, a wartime project in which the Navy supposedly sought to make a ship invisible.[102]

However powerful they could be, the increasing links among science, technology, and progress sometimes implied the loss of something human. Contactees described the space people, who had come to help humans navigate the dangers of the modern scientific age, as beings who utilized advanced science but did not rely solely on it. In almost all accounts, they combined their extensive knowledge of natural laws with spiritual laws and a theology hybridized from Christianity, Spiritualism, Theosophy, and Americanized Eastern religious traditions. Such elements were typically lacking in popular discussion of the later abductees. Their encounters with extraterrestrials often seemed much more frightening and bewildering. Abduction tales typically revolved around detailed medical examinations, and involved UFO occupants who were more often truly alien.[103] A focus on the technological aspects of scientific power among earthlings sometimes seemed to come at the cost of a connection to the human, or even natural, world. Though uttered by earlier Americans, the claim that science appeared to be rushing too far ahead of human wisdom achieved new currency after World War II. While 89 percent of respondents to a 1957 survey on attitudes toward science agreed that scientific research is the "main reason for our rapid progress," 43 percent worried that it "makes our way of life change too fast." The sometimes reckless rush of science into the future manifest itself as a bizarre procession of movie monsters, from the giant insects of *Them!* to Japan-stomping Godzilla, given birth by nuclear science gone awry.

More mundane, but hardly less serious, concerns centered on the impact of science and technology on the natural environment. Elements of the antiestablishment movement of the late 1960s and early 1970s rejected modern technology completely or pushed for the development of alternative and environment-friendly technologies, such as solar power.[104] By the end of the twentieth century, a highly technological science sometimes lost its contact with nature altogether. The phrase "science and nature" proliferated in magazine articles during the 1990s.[105] It also became a standard organizational category in contexts from the board game *Trivial Pursuit,* first introduced in 1983, to news websites such as the one maintained by the British Broadcasting Company. Like previous juxtapositions, such as science and religion, science and nature opened the possibility that the two might involve different kinds of knowledge or meanings. Many present-day advertisers of goods aimed at environment- or health-conscious consumers have, for instance, chosen to market their goods as natural rather than scientific.

The Fragmentation of Method

While depictions of the technological power of science proliferated in popular science talk, other sorts or scientific rhetoric occupied a reduced profile. In decades past, one of the chief rhetorical means of asserting the nontechnological value of science had been through appeals to methodology. But the prominence of method talk among Americans declined and fragmented after mid-century, and science itself frequently became surrounded by discussion of its limitations. The invocation of scientific methodology did not disappear after 1950. The work of German philosopher Karl Popper on falsification first became available in English translation in 1959. Methodological rhetoric, including "the scientific method," also circulated in textbooks through the end of the century. Scientific method remained a powerful keyword in popular debate, as well. Partisans on all sides of the UFO phenomenon declared their adherence to it or their opponents' violation of it. For skeptics, belief in strange ideas like flying saucers "involved rejecting the scientific method and standards of evidence and credibility."[106] Though neither side described the method in detail, UFO enthusiasts obviously disagreed. Still hopeful that the recently announced Condon study might approach the problem of UFOs more seriously than had the Air Force, Hynek celebrated the fact that at last the public would be able "to see the scientific method applied thoroughly."[107] Sadly, such optimism quickly turned sour. One of the most famous methodological quips among UFO advocates in the wake of the Condon report was Hynek's assertion that "ridicule is not part of the scientific method."[108]

Despite its promise, however, methodology lost some of its rhetorical ability to bridge the rhetorical gap between science and the general public. Condon partly blamed scientists themselves for the growing interest in "pseudo-science" because they had "failed rather miserably to give even the so-called educated people some feeling for the way in which science investigates a subject, and the way in which scientists subject their observational material to critical evaluation before reaching conclusions."[109] While Condon may have overstated the case for effect, methodological discussion around the subject of UFOs lacked a focus and coherency present in previous decades. Skeptical scientists and eager ufologists often focused on fairly detailed and specialized topics, such as the Air Force's use of statistics, the means of detecting traces of hoaxing in photographs, or determining the reliability of particular witnesses, with little attempt to make global statements about the way science generally worked. Almost no one sought to transcend local discussions of procedure by citing traditional methodological guides such as Bacon, Newton, or Kepler. The fragmentary use of method talk—focusing on the details of a given photograph to determine its reliability rather than on a general intellectual procedure for evaluating evidence as a whole—suggests the end result of a progressive loss of early-nineteenth-century methodological consensus, if only rhetorical, that at one time bound together those who practiced, consumed, appreciated, and even casually encountered scientific knowledge.

After mid-century, the methodological vocabulary provided by popular books and magazines was narrowing. During the decades after World War II, invocation of the "scientific method," one of the key elements of that vocabulary, circulated less frequently in American magazines than before.[110] It endured increased scrutiny in occasional articles asking, "Is There a Scientific Method?"[111] In 1964, one author suggested that "the scientific method has become a sort of demi-god whose worshippers believe that it is only by this method that truth may be known," though she could not bring herself to entirely reject it, whatever it was.[112] Education historian John Rudolph identified a mid-century movement among scientists and university educators interested in elementary and high school science teaching that targeted "the scientific method" as a stifling and rigid picture of how science worked. Instead, taking advantage of Cold War concerns that the United States might fall behind the

Soviet Union in science, such reformers promoted a far more complex, subtle, and informal method, often depicted simply as exploring the world, which seemed likely to attract the brightest students.[113]

Questioning was paralleled by fragmentation and decline. Though prompting much discussion among particular communities, such as academics, Popper's principle of falsification found little use in magazine discussion of flying saucers. Nor was the stepwise, hypothetico-deductive method that was drilled into generations of prewar students a widespread popular resource. No public commentators on UFOs used the decades-old sequence of hypothesis, experiment, and revision. Instead, students seemed to leave such method talk at the classroom door. Only 10 percent of those asked to give the meaning of scientific study in the 1957 NASW survey mentioned anything approaching such a formula. Instead, when asked to define the nature of scientific study, a full 32 percent did not understand the question.[114] Rather, method often played a secondary role to other kinds of science talk. As the appeals to established science became more widespread, the status of the people making scientific claims often weighed more heavily than the method used. In 1950, a columnist in the *Saturday Review of Literature* differentiated pseudo-science from legitimate science by claiming that "when Velikovsky speaks it is one thing; when Einstein speaks it is another."[115] And although Hynek liked to declare what scientific method was not, he rarely said what it was. Instead, he spent more rhetorical energy on establishing the legitimacy of those who witnessed or studied UFOs than on how reliable conclusions could be drawn from reports. His hope for a demonstration of the scientific method by the Condon committee was based largely on the prominent role of professional university scientists in its ranks.

An institutionalized view of science privileged such evaluations over the more internal, often psychological accounts of discovery that methodological discussion had previously entailed. Moreover, any foray into psychology in the era of flying saucers was dangerous. The psyches of scientists could be subjected to the same scrutiny as those of UFO witnesses and even the hysteria-prone general public. Such critiques were rare, but they did appear in popular literature from time to time. Science journalist Waldemar Kaempffert reminded readers of *Science Digest* in 1947 that "even scientists have been known to 'see things.'" As an example, he offered the identification of Martian canals by the late Percival Lowell.[116] Even more seriously, one author saw fit in *Saturday Review* to repeat the warning of Carl Jung, who took his own stab at explaining UFOs in 1959. Jung claimed that the "super-calm, super-rational type of scientifically oriented person" is also "by very reason of his intellectually narrowed awareness" the sort of human being "most prone to hysteria—in its clinical, rather than its popularly envisaged sense."[117] Clearly, nothing good could come from prying too deeply into the "ravings of the human mind."[118]

In the midst of the atomic and space ages, the technological power of science also frequently obscured its methodology as a primary contribution to modern civilization. The work of scientists as depicted in magazines and popular books often became the operation of advanced equipment, such as computers. Authors of postwar popular fiction, including comic books, tended to portray scientists' work in terms of pushing buttons or changing tape reels on impressive-looking machines. The more intellectual aspects of method often receded. Almost universally, authors wrangling over the value of witness testimony used the phrase "technically trained" synonymously with "scientifically trained." And on numerous occasions, pursuing a scientific study of an area such as UFOs or engaging in "scientific observation" meant simply deploying the "finest technical equipment." One UFO watcher in Texas noted in 1976 that "if we look for ten years with the gear we have now, and don't find anything, . . . then there's nothing there."[119] In some of the most extreme cases, scientists were lost in the technological glare, merged with their equipment. The character of Mr. Spock from the television series *Star Trek*, which began in 1967, suggested just how computerlike the scientist could be. Images of scientists as machines dated back to the late nineteenth century, but they had never before become so widespread or powerful, buttressed by the plethora of high-technology devices that populated the world of the later twentieth century. A close connection to devices also helped to promote an image of objectivity and the ability to overcome the weaknesses of the human mind, which had been the primary purpose of scientific methodology from the beginning. But it accomplished this at the cost of a science with any hint of humanity.

Just as the rhetorical tools available for depicting scientific methodology grew increasingly constrained, so, too, did its assumed application. During the late nineteenth and early twentieth centuries, no shortage of Americans claimed that the scientific method could be exported, with undoubted value, to areas from burying the dead to educating the living. By the late 1900s, such exports of scientific methodology, while not completely gone, were increasingly hard to find. During the 1920s, the bulk of books with "scientific method" in their titles had been about the current and lively topic of social science. After mid-century, increasing numbers of such works dealt with more abstract philosophical issues. Talk about the methods of late-twentieth-century science also became far more tentative than its predecessors. A report on Condon's findings in *Science* pointed to its widespread rejection among advocates as a "reminder that scientific methods are not always able to resolve

problems in fields where emotions run high and data are scarce."[120] It remained
an open question for the historian Ronald Tobey, reviewing a history of UFOs
in America, whether scientific method was "operational in the theater of a
non-scientific public" at all.[121]

The tentativeness of some method talk used in debates over UFOs could
be, and sometimes was, used to construct a restricted view of science. Many
Americans did believe that science had its limits. While 38 percent of respondents to the 1957 NASW survey believed that science "may understand most
things," 28 percent claimed it "will never understand a lot of things," and
18 percent thought it would "not understand everything."[122] UFOs provided
one example of something possibly outside the reach of science. By the early 1970s, ufologists such as Vallee and Hynek had begun openly to question
whether the methods of science were applicable to the UFO phenomenon
in any form. Similar questioning continued into the closing decade of the
century. Vivienne Simon, executive director of the Center for Psychological
and Social Change and a supporter of the reality of alien abduction, told a
Time magazine reporter that the goal of her organization was to "challenge the
current scientific method, which is to deny all things you cannot reduce to
statistics." Other UFO researchers echoed the views of some late-twentieth-
century Americans who expressed admiration for non-Western societies, free
from the crushing weight of science and technology. One prominent ufologist
lamented in 1994 that "we have lost the faculties to know other realities that
other cultures can still know. . . . We've lost all that ability to know a world
beyond the physical."[123]

God Drives a Flying Saucer

Despite the numerous images of a limited scientific enterprise that circulated
in popular literature, many of those engaged in the debate over UFOs sought
to depict science as something more than a producer of gizmos and gadgets.
Such attempts sometimes led into dangerous territory. In the absence of a vital and engaging repertoire of methodological rhetoric, authors in search of a
broadly relevant and important science often expressed themselves in religious
terms. Religion, as the most compelling and important example of nonscience
for a wide variety of Americans, provided a set of images, ideas, and language
that could be used to bridge the perceived gap between science and the rest
of culture. Some devotees of relativity, as we have seen, had constructed similar bridges during the 1920s and 1930s. But their crossing had already become
plagued with tension in the decades before World War II. By the last third of
the century, it had become particularly treacherous. The mixture of science

and religion became further associated with marginal areas that smacked of some transgression such as pseudo-science. In reaction to the antiestablishment movements of the late 1960s and 1970s, which often turned against both orthodox science and orthodox religion in the quest for something broader than either one, links between scientific and religious terrain were sometimes cut from both sides, an action that left many of those seeking expansive images of science little option but to turn away from orthodoxy completely.

During the 1950s, it was still possible to construct depictions of a religiously infused science without serious eruptions of negative public rhetoric. Historian James Gilbert has discussed a variety of attempts to harmonize science and religion in the decade and a half after World War II, from the use of evangelical science films by military chaplains to a series of television shows by director Frank Capra discussing scientific topics in strongly religious terms. Some science fiction of the era also mixed the scientific and religious. The film *The Day the Earth Stood Still* told the story of an extraterrestrial visitor named "Mr. Carpenter" (an allusion to the original occupation of Jesus) who came in an attempt to establish world peace, died at the hands of a fearful military, then was resurrected before ascending into heaven in his flying saucer. Tension sometimes accompanied such efforts, as in the frequent debate between Capra and the scientific advisors, who often objected to the religious tone of his scripts. But that tone did not become the occasion of widespread teeth gnashing or debate.[124]

Nevertheless, depictions of science infused with religious relevance did not dominate public discussion, even amidst debate over UFOs. Instead, the invocation of religious rhetoric appeared only among fringe groups, with little interest in mainstream recognition. The contactee movement merged scientific jargon and religious mysticism in ways that struck most outside observers, such as NICAP and ufologists striving for orthodox recognition, as illegitimate. Rather, UFOs were a problem of national defense, if they were a problem at all. Nor were most Americans queried by the 1957 NASW survey inclined to mix their science and religion too completely, let alone adopt a "religion of science." When asked whether science or faith should primarily guide people, 50 percent chose faith. Only 21 percent thought science should predominate, and a mere 12 percent advocated a combination of both.[125] Science was often depicted as valuable for its presumed medical and technological improvements and for its role in protecting the nation instead of its intellectual contributions to human civilization. Conflict over attempts to harmonize science and religion would have done little to aid a scientific community already experiencing unprecedented levels of professional security and support. In the somewhat conservative atmosphere of the 1950s, such conflict might even have reflected badly on scientists.

By the late 1960s, however, UFOs increasingly appeared in a religious light. This turn toward religion language paralleled the discontentment of many UFO advocates with established science, which rose to a head at the end of the decade, and the subsequent hope for a new and expanded scientific endeavor. UFOs, in their opinion, provided one of the keys to bringing about a "mighty quantum jump" in views about the universe and served as a means of enacting one of science fiction author Arthur C. Clarke's many dictums: "[T]he only way to define the limits of the possible is by going beyond them into the impossible." Such attitudes only intensified as many ufologists realized that the Condon report had frozen them out of the orthodox scientific establishment, and they found themselves relying on public support more than ever. In 1973, Vallee lauded Hynek as someone "who happened to combine an astronomer's passion for the discovery of scientific truth with an almost mystical insight into its limitations." Vallee became committed to working to find a single cause behind both UFO sightings and reported miracles such as the appearance of the Virgin at Fatima. During the 1970s, many others interested in UFOs turned away from the extraterrestrial hypothesis and toward psychic, mystical, or paranormal explanations in the search for a "reality beyond reality." One witness sought to describe his UFO encounter by invoking what arguably became the mascot of the alternative movement of the late 1960s and 1970s, asking "'Have you ever been close to a whale?'"[126]

The mixture of scientific and religious rhetoric was most evident in what had once been out of bounds for any ufologist hoping for orthodox recognition, namely, the discussion of extraterrestrial beings. The very existence of such beings as the ultimate outsiders often came to symbolize the hoped-for expansion of science. Descriptions of their presumably advanced science and technology, a vision perhaps of what ours would become, frequently evoked adjectives such as "god-like." In the hands of the contactees, Venusians and Martians themselves took on a strongly Messianic air. Adamski and others claimed that Jesus had been a space brother sent to earth to help humanity. Later authors, including Erich von Daniken, whose *Chariots of the Gods* (1968) had sold a million copies by 1974, and Connecticut schoolteacher R. L. Dione, who penned the slightly more obscure *God Drives a Flying Saucer* (1968), continued the tradition of locating aliens at the root of religious mythology. Images of angelic aliens also appeared in films such as 1978's *Close Encounters of the Third Kind*.[127] Though many accounts of ancient or angelic visitations supposedly involved material beings, at other times, extraterrestrials straddled the boundary between the physical and nonphysical worlds. N. Mead Layne's San Diego–based Borderland Sciences Research Associates tended to view flying saucers as the ethereal craft of equally ethereal visitors.[128] A letter to

Look by a reader in Casper, Wyoming, suggested that "materialistic people who can't believe in spiritual or astral planes of existence have been badly deceived."[129] Eventually, most ufologists moved toward less materialistic ideas, either linking the appearance of UFOs to extradimensional entities or to the same psychic forces responsible for miraculous events.

Meanwhile, movements that detractors typically identified as pseudo-science, such as those associated with UFOs or creation science, increasingly pushed orthodox critics to portray the mixture of science and religion as illegitimate. The boundary between science and religion was sometimes enforced by that ultimate arbiter of American vocabulary, the dictionary. A 1966 article in *Time* informed its readers that "religion, by one Webster definition, is the object of a pursuit arousing 'religious conviction and feelings such as great faith, devotion, or fervor,' and science, by another Webster definition, is 'accumulated and accepted knowledge which has been systematized.'" The two merged "in the field of UFO where the writers on the subject certainly show great faith, devotion and fervor in their efforts to have the objects regarded as part of accepted and accumulated knowledge." While not immediately critical of such a combination, the author went on to suggest that Frank Edward's "mixture of science and religion is curious, as if Billy Sunday had undertaken a sermon on the subject of the binomial theorem."[130] Other mixtures of science and religion caused greater discomfort. A reviewer of the work of Gerald Heard, the author of an early UFO tome called *Is Another World Watching?* (1951), expressed immediate suspicion about Heard's motives and ability to maintain an open mind, based on the subject matter of his previous books, including *A Preface to Prayer* (1944) and *Is God in History?* (1950).[131]

Skeptics and doubters routinely referred to those who accepted the reality of UFOs, and particularly contactees, as "believers," "flying-saucer cultists," or as adherents to a "sci-fi religion." Carl Sagan, among others, claimed to detect "unfulfilled religious needs" behind the UFO epidemic. The "distant and exotic worlds and their pseudo-scientific overlay" of UFO advocates appealed, Sagan speculated, "to many people who reject older religious frameworks."[132] But representatives of those older frameworks often regarded UFOs with little more enthusiasm than the self-appointed defenders of science. From this perspective of mainstream religion, the UFO phenomenon could appear to be as much "pseudo-religion" as pseudo-science. In 1978, Irving Hexam, an evangelical philosopher of religion, expressed concern that Christian book publishers were printing too many works on topics such as UFOs and the Bermuda Triangle, which were full of "pseudo-scientific and semi-occultist" ideas.[133]

The eventual decay of bridges between the scientific and religious, partly under the weight of so many attempts to cross them, appeared in the word choice of Americans. Use of the phrase "science and religion" not only declined in the titles of magazines over the last half of the 1900s, but was also eclipsed by references to the more relevant and far safer juxtaposition of "science and technology."[134] For those seeking a scientific enterprise that could encompass the whole realm of human experience and offer a grand view of the universe, orthodox science increasingly seemed the wrong place to turn. For some former devotees, science itself was becoming increasingly narrow. At the very end of his *Invisible College* (1975) a somewhat pessimistic Jacques Vallee reflected that "for a long time I believed that science would gradually realize the importance of paranormal phenomena as an opportunity to expand its theories of the world. . . . Now I believe differently. . . . It is no longer to science that we must turn," he counseled his readers. Instead "the solution lies where it has always been: within ourselves."[135]

Weird Science

Overall, then, the vast majority of rhetorical tools Americans used in the debate over UFOs worked to inscribe clear and robust boundaries around science. A wide variety of depictions presented scientists as a distinct group of people who worked in a set of institutions that were largely closed to outsiders. The science they wielded, by most accounts, was capable of achieving amazing feats. But the words, phrases, and ideas widely available in popular books and magazines could also construct a very distant science. The same rhetorical tools through which science received its power often strained the connections between the scientific enterprise and other areas of culture. The severing of links between science and the world of ordinary people reached its zenith in concern about secrecy, but it was reflected elsewhere, too, in the decline and fragmentation of once important areas of science talk. Invocations of methodology played a relatively minor role in debate over UFOs. Attempts to expand the limits of science, often by appeals to religious language, faltered amidst accusations of transgression. By the last third of the 1900s, the only unproblematic link between science and the lives of most people came through technology, a situation that did not please everyone. "Even in the 1980s," John Burnham conceded, "there were still a few men and women of science who had a sense of identity in which science was a way to truth, civilization, morality, and other constructive value." But such people were "swamped" by colleagues who were mere technicians laboring without true zeal or faith in science.[136]

Still, it was possible even for late-nineteenth-century Americans to construct images of science that resonated with personal and intellectual, rather than just technological, power and relevance. Skeptics often presented science as a central part of their lives, even though to outsiders they may have looked narrow and rigid. Many advocates of UFOs also sought to make science meaningful, not by adjusting themselves to the seemingly pared-down universe of orthodox science but by reconstructing the scientific. Americans had publicly sought truths beyond science well before the final half of the twentieth century. Others had claimed "true" science on behalf of their own beliefs. But another option became particularly visible in popular discussion of science after the 1960s. It occupied a middle ground between confining science to a narrow range of phenomena and methods and redefining it in ways that included one's cherished ideas; rather, some Americans constructed an "alternative" science. Even Vallee's rejection, in the face of frustration at ufologists' struggle to make UFOs scientific, was not total. In 1992, he published his journals from the late 1950s through mid-1960s under the title *Forbidden Science*.[137] The rhetorical formation of an alternative scientific enterprise testified to the strength of the boundaries around science by the last third of the twentieth century. But that formation also implied the otherness of science, its inaccessibility, and even irrelevance. Instead of turning to scientific knowledge or methods, many of those, such as Vallee, who wanted to discover some greater meaning in science found the need to transcend it and thus constructed their own, self-consciously alternative body of knowledge about the world.

Many skeptical observers referred to this body of knowledge as "pseudo-science," a term that not only became more widely used after the late 1960s but increasingly came to mean something distinct. Before the mid-twentieth century, most authors had associated it with mistaken beliefs that masqueraded as scientific knowledge, but there was little attempt to link these beliefs into a coherent whole. In 1932, H. L. Mencken suggested that each branch of science was shadowed by an evil twin, "a grotesque Doppelganger."[138] But he did not gather these individual doppelgangers into the more general category of pseudo-science. By the third quarter of the 1900s, critical scientists and popularizers depicted growing numbers of linkages among a variety of strange topics, including "everything from PK (psychokinesis, moving things by will power) and astral projection (mental journeys to remote celestial bodies) to extraterrestrial space vehicles manned by web-footed crews, pyramid power, dowsing, astrology, the Bermuda triangle, psychic plants, exorcism and so on and so on."[139] Challenging the hydra of a distinct pseudo-science emerged as a major concern of CSICOP, of which Sagan and many other science popularizers were members. The *Skeptical Inquirer* provided one of the most important sources of elaboration and exploration of a reified concept of the pseudo-scientific.[140]

The notion of pseudo-science that resulted from such rhetorical work was more than simply a term to be applied to mistaken beliefs or garbled science. It was an alternate system of belief, a doppelganger of science as a whole. Where science was potentially unpopular, pseudo-science was inherently popular. Since the late nineteenth century, some Americans had claimed that true, serious science fared poorly in a marketplace of ideas increasingly dominated by commercial concerns and in which entertainment value predominated over truth. Hynek complained in 1968 that nothing would help the cause of ufology more than "a heavily footnoted, scientific document, painstakingly probing into every crack and corner to establish incontrovertibly that even one true UFO exists." However, such a book "wouldn't sell."[141] Pseudo-science was also frequently depicted as incorporating a religious, and thus archetypically nonscientific component. As we have seen, Sagan attributed belief in pseudo-scientific topics to an unfulfilled religious need. Likewise, pseudo-science could imply the reverse of scientific methodology, a substitution of faith for reason, or even a rejection of proper method entirely. Sagan further claimed the reason some people turned toward UFOs or astrology "is precisely that they are often beyond the pale of established science, that they often outrage conservative scientists, and that they seem to deny the scientific method."[142]

Pseudo-science was intended for rhetorical battle and only imperfectly overlapped what Vallee and those like him would have called forbidden, alternative, or unorthodox science. But it did capture something that was actually happening among advocates of UFOs, as well as other sorts of unusual phenomena. Just as a coherent vision of pseudo-science had been anticipated before World War II, the groundwork of an alternative science was laid during the 1920s and 1930s. The most influential source of inspiration among those interested in alternative science was found in the work of former newspaper reporter Charles Fort. Mencken's statements about doppelgangers were from a review of Fort's book *Wild Talents* (1932). In this and several other tomes, the first entitled *The Book of the Damned* (1919), Fort sought to parade in front of the reader all of the phenomena, ignored and dismissed by modern science, that he called the damned. Such dismissal was not simply a feature of the modern incarnation of science, but was a necessary feature of science. Even in a different form, it would need to exclude something.[143] Among such excluded phenomena, he discussed strange objects seen in the sky, rains of frogs and other unusual things, hauntings, spontaneous human combustion, and a host of other seemingly inexplicable events. Though more or less unknown among the public at large, as noted above, Fort inspired the creation of the Fortean Society in 1932. Author Tiffany Thayer, the driving force behind the group's formation, characterized the society as the "Red Cross of the human mind" and listed as one of its major goals "to perpetuate dissent."[144] Fort's work prompted the publication of occasional Fortean magazines, which served as compendiums of unusual phenomena, and was frequently cited as an inspiration by a large number of UFO researchers, including Vallee.

The scope of this increasingly articulated alternative science was vast. If the *Skeptical Inquirer* provided the raw materials for constructing a new concept of pseudo-science, *Fate* provided a forum where columnists and correspondents worked out the general outlines of alternative science. *Fate*'s August 1971 issue included articles on an appearance of the Virgin Mary in Egypt, psychic detectives in Arizona, a ghost sighting, and UFOs. But *Fate* was not alone in gathering together such a wide array of phenomena. As mentioned above, by the 1960s, Vallee had begun to link miraculous visions with UFO encounters. John Keel, the author of several books during the 1960s and 1970s, including *The Mothman Prophecies* (1975), went much further in suggesting that a single mysterious phenomenon lay behind a whole range of strange events, from UFOs to the appearances of monsters such as the Mothman.[145] Claims of correspondence between Bigfoot and UFO sightings became a common feature in the literature. A series of television shows, from *In Search of . . .* during the 1970s to *The X-Files* during the 1990s, effectively demonstrated the conglomeration of unusual topics, tackling the question of psychic phenomena one week and the Bermuda Triangle the next. Perhaps the most popular vehicle for this congealing body of knowledge was the supermarket tabloid, which happily ran stories about the discovery of cities on Mars next to features on the bat-faced boy.

There seemed little rationale for connecting these subjects except for the fact that they were strange—that is, rejected by what was sometimes called "conventional science."[146] Indeed, a brief search of the terms "weird science" and "strange science" on the Internet—terms not used pejoratively by those interested in or amused by them—will reveal content very similar to that of *Fate*. Two of *Fate*'s three sections were entitled "True Reports on the Strange and Unknown" and "News and Notes on Unusual Topics." Frank Edwards also presented an expanded view of mysterious phenomena in *Stranger Than Science* (1959) and *Strange World* (1969). Fascination with the unusual was, of course, nothing new; P. T. Barnum exploited it during the nineteenth century. But it had not been pursued with quite so much vigor or so much desire to make a genuine, if highly diffuse, body of knowledge as after World War II, and especially after the late 1960s. By the 1970s, the subject of that knowledge often came to be called the "paranormal" or sometimes the "supranormal," both of which went out of their way to distinguish the events being discussed from the ordinary—despite the fact that many of its advocates claimed paranormal events happened as regularly as normal ones. Perhaps the most obvious effects of the development of a culture of the strange in America has been the frequent appearance of the letter X to denote something unknown and beyond the usual, and the larger idolization of the "extreme," from which its usage has often emerged. Fort's first attempt at writing about his ideas was titled X.[147]

The often cheerful acceptance of a position outside of science, even for those intent on making serious statements about the material world, suggests the extent to which ideas about the scientific endeavor had constricted by the late twentieth century. Where earlier generations wrangled over the name of science, by the last quarter of the twentieth century many partisans in the debate over UFOs took a much more ambiguous stance. At times, articles and advertisements in *Fate* did invoke science to defend their ideas, but at other times modern science played the part of a stodgy authority, sometimes linked to the "Establishment," the transgression of which was a badge of honor. It is notable that many of those who spoke for alternative knowledge rarely claimed the label of scientist for themselves. Such an ambiguous approach to science also appeared in the development of "alternative" medicine. Ufologists among others did seek to legitimate the phenomena that they studied and to bring them into the fold of the known. But by the late 1960s, they represented only a small part of a much larger cultural phenomenon that, in many ways, refused to be confined. For such alternative bodies of knowledge to survive, they had to defy all attempts at explanation. At the very least, most post-Condon ufologists had come to believe that the legitimization of UFOs would require revolutionary changes in the practice of science itself. During the mid-1970s, even Hynek began discussing a characteristic of encounters he called "strangeness."[148]

Ultimately, an alternative science sought to re-infuse the scientific with importance and transcendent value by returning it to an earlier, almost nineteenth-century, form. Self-consciously unorthodox science deeply imbibed a stress on diffusion that would have made any early-nineteenth-century itinerant phrenologist proud. It spread not only through paperback books, but also in pulp magazines, such as *Amazing Stories* or *Weird Tales*, and eventually in the supermarket tabloids. Likewise, scholars have sometimes depicted pseudo-science as arising to "fill the gap between orthodox science and the need for meaning no longer provided by traditional religion."[149] Adamski blasted modern physical science for its emphasis on "'hows' and 'whats,'" a tendency that could be cured by turning toward "[a]rcane science," which "seeks ever to penetrate towards the ultimate, absolute WHY."[150] More accurately, advocates of alternative science often strove to reoccupy territory abandoned by depictions of a distant and well-bounded science, to bring scientific knowledge and practice back within the fold of a harmonious world of truth. The amalgamation of scientific and religious rhetoric provided one means of doing so. An advertisement in Fate showed Jesus, Buddha, and Virachocha of Peru. The caption below claimed that "these three teachers were the world's first nuclear physicists." In exactly what way that was so, the text of the advertisement left unclear, though it did later equate "metaphysicists" with nuclear physicists.[151] Finally, the method of Fort and others represented what earlier generations would have called induction, though partisans in the debate over UFOs seemed to have lost that terminology. In his several books, Fort tirelessly collected bits and pieces of data on tens of thousands of small note cards, looking for patterns.[152] Anyone could take this approach, and the practice of alternative science was generally depicted as far more open than its orthodox partner. Such a science also seemed far more connectable to ordinary human beings, as in attempts by New Agers to construct what one observer has called a "kinder, gentler science."[153]

Still, most images of science in popular books and magazines discussing UFOs were very distinct, typically powerful, but also remote. That remoteness, which had become inherent in American science talk, was exacerbated by the declining availability of many rhetorical tools that had allowed previous generations to express their ideas about science. The need for Americans to search out their own means of making science relevant contributed to the fragmentation of science talk, as different special interest groups pursued their own depictions of science for their own purposes. Most Americans, meanwhile, probably gave little thought to the nature of science. The rhetorical walls around it insulated its practitioners by and large from serious public interference, save when they sought to extend their science into areas marked by ethical or religious controversy, while leaving a one-way technological bridge over which passed intellectually neutral gifts to insure the good will of the masses. Those who saw themselves as outside the scientific community possessed the tools they needed to keep science at arm's length if they wanted, to limit its power over their own lives in the here and now, and to justify their relative ignorance of its content. Thus in late-twentieth-century American popular culture, science had become both incredibly powerful and eminently ignorable.

UFOs, meanwhile, refused to go away. They returned to prominence during the 1990s, partly due to the 1994 rediscovery of the story of a crashed flying saucer near Roswell, New Mexico. Elsewhere, tales of the unusual goings on at the Air Force test facilities in the mysterious "Area 51" were linked to captured alien technology.[154] Likewise, though the contactee movement withered after the 1950s, its mythos continued to exert an influence on the UFO phenomenon, particularly among UFO cults that proliferated during the

1970s. This tradition reemerged forcefully and tragically into the public eye with the 1997 suicides of members of Heaven's Gate, a group that believed that they were part of an "away team" sent from another world.[155] Finally, the 1990s were a boom time for tales of alien abduction from sources such as Harvard psychiatrist John Mack, historian David Jacobs, and authors Budd Hopkins and Whitley Strieber.[156]

As the twenty-first century dawned, however, UFOs ceased to capture quite so much public attention or generate huge quantities of science talk. The terrorist attacks in 2001 may have helped to shift attention to more terrestrial matters, though there have been occasional claims of UFO sightings near the World Trade Center on the morning of September 11. A few groups, particularly a Canadian movement called the Raelians, have also carried on the tradition of the contactees/abductees. The Raelians were organized around the teachings of Rael, a journalist who claimed to have been visited by an extraterrestrial messenger. The Raelians injected themselves into a surprising number of important late-twentieth- and early-twenty-first-century scientific discussions, including an unproven claim that they had successfully cloned a human being. They also supported the notion that aliens had played a role in the evolution of earthly life.[157] This was not a radically new suggestion. The numerous books of Erich von Daniken presented mounds of evidence during the 1970s for similar claims about the involvement of extraterrestrial visitors in the emergence of human civilizations around the globe.[158] However, by the early 2000s, discussion of the possibilities of otherworldly and perhaps extrascientific interventions into evolution, human and otherwise, were dominated by debate over so-called intelligent design rather than UFOs.

Chapter 5   Intelligent Design

The Evolution of Science Talk

DURING THE SUMMER AND fall of 2005, the meaning of science became a topic of intense public controversy in America's heartland. That controversy began when, in order to introduce more "objectivity" in the teaching of Darwinian evolution, several Kansas school board members proposed revising the definition of science in state education standards.[1] Rather than follow the previous standards in stressing the restriction of scientific knowledge to purely natural explanations of the world, the new version focused on the methodological aspects of science, including "observation, hypothesis testing, measurement, experimentation, logical argument, and theory-building."[2] The difference was subtle but it was enough to leave the door open to nonmaterialistic accounts of the history of life that many critics claimed were beyond proper scientific boundaries. A review of the proposed changes by Robert Dennison, formerly president of the Texas Association of Biology Teachers, charged that proponents were trying to "alter the very definition of science."[3] Keith Miller, an assistant professor of geology at Kansas State University, expressed similar concern over what he saw as "a political effort to force a change in the content and nature of science itself."[4]

Debate over the science standards in Kansas formed one front in a national struggle over the scientific status of so-called "intelligent design" or ID for short. ID was founded on the self-styled scientific claim that biological organisms showed evidence of potentially supernatural engineering beyond the reach of random mutations and natural selection. Since the 1990s, its advocates had been promoting their ideas, organizing themselves, and participating in attempts to modify the teaching of evolution in public schools in Ohio, Pennsylvania, Washington, and a number of other states. I do not intend to enter into an in-depth examination of the ID movement here. Once passions have cooled, it will surely provide a rich source of social and cultural insights. But as yet, even in light of a December 2005 federal court decision barring it from classrooms in Dover, Pennsylvania, the story of ID is still being written. Rather than engage in a preemptive history, I will be using the labors of its supporters and the determined work of its critics as a means of identifying and clarifying major themes that have emerged from the last two centuries of American science talk.

A number of intellectual threads connect turn-of-the-twenty-first-century discussion of ID with the episodes I have already examined. Most obviously, claims about the design of living things fundamentally involved the scope and sufficiency of natural selection in particular and naturalistic evolution in general. In making or disputing these claims, partisans reopened issues that were vigorously debated during the nineteenth century, often in almost identical terms. Some supporters of ID also extended questions about the limits of naturalism to the fundamental details of the physical universe and its origin. The Big Bang predicted by the field equations of general relativity provided one potential barrier to the extension of purely natural causes, just as relativity itself struck some early-twentieth-century admirers as a bulwark against an overly materialistic science. Finally, ID involved one of the same issues that was at the heart of controversy over UFOs, namely, the means by which observers could distinguish between purely natural phenomena, whether ball lightning or the outcomes of natural selection, and the products of conscious and intelligent intention. Though most advocates of ID suggested that God was responsible for the engineering of modern life, some admitted that it might have also been the work of visiting extraterrestrials.

Rhetorically too, the science talk that has so far permeated and framed discussion of ID fed from the same streams that flowed through the debates over phrenology, evolution, relativity, and UFOs. Most depictions of science built on the well-bounded foundations established during the nineteenth century. Back and forth regarding naturalism as a defining aspect of scientific knowledge became one aspect of a much larger struggle over the distinctions between science and religion. At other times, public commentators on all sides turned to additional contrasts between science and what it was supposedly not, such as philosophy or pseudo-science. As the controversy over the science standards in Kansas indicated, many of these boundaries were drawn by reference to the unique methods of science. And invocations of the division between scientists and the general public and between scientific and ordinary knowledge circulated widely. These assertions were shadowed, meanwhile, by the same kinds of discomfort with strict scientific boundaries that frequently rose to the surface during the twentieth century. Critics sometimes charged that ID was both a "science-stopper," in that it called a halt to the scientific investigation of the origins of natural phenomena, and an illegitimate expansion of science into the domain of supernatural phenomena. Neither of these was completely accurate. Yet together they caught something of the ambivalent role of science in the rhetoric of ID. Like those seriously interested in UFOs, ID's supporters celebrated science, pointed out its limits, and sought to create a self-consciously alternative, more easily accessible, and deeply meaningful scientific enterprise.

Designer Science

ID emerged at the end of a succession of twentieth-century social movements that coalesced around rejection of evolutionary accounts of life, especially in public classrooms. The antievolution crusade of the 1920s was followed, beginning in the 1970s, by attempts to promote "creation science," which sought not only to disprove biological evolution but also to put the biblical story of the fairly recent and divine creation of the earth in six literal days on scientific footing. Some individuals interested in creation science organized themselves into institutes, where they tried to foster a research program based on their ideas, though there were occasional clashes between those seeking to meet the standards of professional academic scientific work and those who were less concerned with such expectations. On the broader public stage, activists seeking to resist the teaching of evolution in public schools shifted from working to outlaw such teaching to promoting equal time in science classes for scientific creationism alongside evolutionary ideas. By the 1980s, however, creation science underwent a series of legal tests in Arkansas, Louisiana, and finally the U.S. Supreme Court, which declared it to be religion rather than science, and thus deprived it of any place in public educational institutions.[5]

The first hints of another path toward scientifically defeating evolution appeared just as creation science was meeting defeats of its own in state and federal courtrooms. In 1985, a British publisher released a book by Michael Denton called *Evolution: A Theory in Crisis*. An American version appeared the following year. Denton was a physician who had earned a Ph.D. in biochemistry at London University during the early 1970s. As he completed his doctoral work, he also began to experience severe doubts about Darwinian evolution. His response was not to turn to creation science or to consult the book of Genesis; *Evolution* contained no statements about the Creator or His methods and avoided allusions to Scripture. Instead, Denton self-consciously subjected evolution by random mutation and natural selection to what he depicted as a strict scientific evaluation. His conclusion—that Darwinian evolution was not supported by empirical evidence or even plausible speculation and could not account for biological change beyond minor modifications within species—moved very few members of the orthodox scientific community. Rather, a variety of biologists strongly criticized Denton's book.[6] Nevertheless, Denton's efforts did eventually inspire a number of figures that later played foundational roles in the emergence of ID during the next decade.

One of these figures was University of California–Berkeley law professor Phillip Johnson, who assembled his own case against Darwinian evolution in *Darwin on Trial* (1991). Like Denton, Johnson did not appeal to the Bible or draw conclusions about the nature of whatever force was ultimately responsible for earthly life. Denton's work also influenced Michael Behe, a professor of biochemistry at Lehigh University. Behe became one of the leading lights of what was becoming more widely known as "intelligent design" with the publication of *Darwin's Black Box* (1996). Behe expressed a willingness to accept the ability of Darwinian natural selection to explain huge changes in biological organisms and did not reject the notion that disparate species had common ancestors. Instead, he focused on the microscopic level, on what he routinely called the biochemical "machinery" of life. In *Darwin's Black Box*, Behe presented a number of such mechanisms, including the bacterial flagellum, that he claimed were "irreducibly complex," by which he meant that the whole could not function without all of its parts assembled just so. A partial flagellum was of no value and thus the slow, step-by-step process of natural selection acting on tiny changes could not have produced the completed mechanism. The presence of irreducibly complex structures in living things suggested for Behe the work of some designing intelligence, though he followed the lead of Denton and Johnson in saying little about who or what he thought this intelligence might be. A few years later, William Dembski, who had recently augmented his Ph.D. in mathematics with a master's degree in theology and another Ph.D. in philosophy, joined Behe as an intellectual prime mover of intelligent design. Rather than approaching the problem from biochemistry, Dembski discussed it in terms of information and sought to provide mathematical tools that could test for the presence of intelligent engineering in complex systems, such as biological organisms.

Over the 1990s, the banner of ID was also being taken up by a loosely affiliated collection of groups and individuals who shared a desire to oppose evolution. Some had roots in more traditional creation science. Others hailed generally from the right wing of late-twentieth-century social, political, and intellectual thought. Advocates and critics of ID routinely depicted it as representing one front in the so-called "culture wars," a term coined in 1990 to describe perceived struggle between liberal and conservative elements of American society. By some accounts, this struggle had been ongoing since the 1960s, though by the 1980s right-leaning countercultural movements were more active in publicly opposing orthodoxy than left-leaning ones. For instance, a number of ID enthusiasts explicitly linked belief in Darwinian evolution and the materialism that supposedly attended it to such "far-reaching consequences for Western society" as the erosion of "traditional theories of human freedom and responsibility" and the rise of "moral relativism."[7] ID also found an important institutional home in the Seattle-based Discovery Institute, a think tank founded in 1990 to express a range of beliefs in, among other things, "God-given reason and the permanency of human nature," "free market economics domestically and internationally," and "the potential of science and technology to promote an improved future for individuals, families and communities."[8] In 1996, the institute established the Center for the Renewal of Science and Culture (later renamed the Center for Science and Culture) to cultivate work on ID, organize conferences and seminars on topics related to it, and provide an institutional affiliation for its leading figures, including Behe, Dembski, and Johnson.

ID also met with stiff resistance from a variety of scientists, philosophers, journalists, and educators. Though many scientific practitioners were initially hesitant to fully engage advocates of ID in public, a few, including the well-known biologist Stephen Jay Gould and geologist Kenneth Miller, challenged their claims in print and in interviews. Philosopher Robert Pennock, who argued against what he called the "new creationism" in his *Tower of Babel* (1999), and Eugenie Scott, executive director of the National Center for Science Education, likewise took up prominent positions defending Darwinian evolution and strongly criticizing ID. Where controversy over teaching Darwinian evolution in schools erupted, partisans organized themselves into rival groups. In Kansas, the Intelligent Design Network faced off against the pro-evolution Kansas Citizens for Science. Critics also linked ID to a set of larger issues, from the preservation of Enlightenment values to the resistance to "anti-science," "irrationalism," or the supposed attempts of some right-wing Christians to turn the United States into a "theocracy."[9] The trial sparked by ID-inspired criticism of evolution in Dover, Pennsylvania, schools was initiated by the American Civil Liberties Union, widely depicted as a major combatant in the culture wars. A common tactic in the struggle against ID involved placing it beyond the boundaries of legitimate science, a move that both helped to delegitimize the claims of its supporters and provide a ready-made case, in light of the separation of church and state, to exclude those claims from public classrooms. In *Creationism's Trojan Horse* (2004), philosopher Barbara Forrest and biologist Paul R. Gross took pains to make clear that ID was "everything *except* science."[10]

Scientific IDs

Whatever else divided them, admirers and critics of ID almost universally talked about science in a strongly bounded way, as if it were clearly separate from other kinds of knowledge and, at least in principle, distinguishable from them. Defenders of orthodox evolution appealed widely to a distinct scientific community as the ultimate arbiters of what was science and what was not. In a letter arguing that the burden of proof in the case of ID was on its proponents rather than on scientific researchers, Michael Friedlander, professor of physics at Washington University, claimed in no uncertain terms that "the content of science is decided by scientists alone."[11] Rather than dispute such assertions, advocates of ID routinely sought to demonstrate their membership in the exclusive club of established science. More than adherents to any previous antievolutionary movement, ID enthusiasts stressed the orthodox credentials and institutional affiliations of their leading thinkers. In a response to *Tower of Babel*, Behe claimed that instead of the "creationists" he railed against, Robert Pennock was really facing opponents who "turn out to have doctorates in embryology, biochemistry, the philosophy of science, and mathematics from places like the University of Chicago, Cambridge, and Berkeley."[12] Other supporters of ID called attention to a list of "400 scientists" compiled by the Discovery Institute, "including those from such prestigious institutions as Princeton, MIT and Cornell," who reportedly expressed skepticism over Darwinian evolution.[13] Citations of character or intimate knowledge without the official sanction of academic institutions were, by and large, nowhere to be found.

There was also considerable controversy over the presence of ID specifically in peer-reviewed scientific publications. In *Creationism's Trojan Horse*, Forrest and Gross noted that Behe had "published no scientific, peer-reviewed research on any aspect of ID in any scientific journal."[14] Such lack was alternately contrasted with the overwhelming presence of evolution in modern scientific literature. An article by the editor of the *Skeptical Inquirer* asserted that ID had been "refuted by peer-reviewed science. So let's hear no more about Dr. Behe."[15] These depictions offered the institution of peer review as a surrogate for the scientific community as a whole. Advocates of ID, by contrast, claimed that their leading researchers had in fact "presented evidence in peer-reviewed books published by major, prestigious publishers and in peer-reviewed articles published by major, prestigious journals."[16] Meanwhile, Behe reported a survey of major peer-reviewed journals for what he considered solid work on molecular evolution and found little to speak of, a result he interpreted as indicating something insufficient in current notions of Darwinian evolution on the microscale.[17]

Discussion of ID also drew on the methodological vocabularies of Americans. The degree to which it did so suggested a reinvigoration of method as a prominent aspect of science talk compared with debate over UFOs, often through the use of rhetorical tools borrowed from the philosophy of science. It was still possible to ask, echoing questions raised during the 1950s and 1960s, whether there was any definite scientific method at all. And it was still possible to answer no. Behe and Dembski both quoted Percy Bridgeman's quip that "the scientific method, as far as it is a method, is nothing more than doing one's damnedest with one's mind, no holds barred." For added emphasis, Dembski cited philosopher Paul Feyerabend's *Against Method* (1975), which denied outright the existence of a formal scientific methodology. By and large, however, such questioning was rare. It appeared that talk about a uniquely scientific method was too useful to abandon. In his letter to the *Chronicle of Higher Education*, Michael Friedlander further appealed to his colleagues to "devote some of our class time to discussing the methods of science, not only its content."[18] Yet the vast majority of the methodological terms and ideas made visible by ID did little more to make scientific thinking accessible or show how it might be extended to a wide range of contemporary issues than mid-twentieth-century scientific rhetoric. Instead, talk about methodology continued to act primarily as a gatekeeper, exorcising outsiders, protecting insiders, and distinguishing uniquely scientific from other kinds of knowledge.

Those who rejected ID frequently stressed that proper scientific method demanded "experiments whose results can be measured and replicated."[19] ID was ultimately "untestable," champions of evolution charged, and therefore unscientific. Some methodological critiques, far more than amidst debate over UFOs, made reference in particular to Karl Popper's falsifiability, a move inspired partly by the use of Popper's work in the court cases against creation science during the 1980s and by the substantial involvement of philosophers of science in the controversy over ID. In response, those defending ID stressed that the ability to detect complex design in organisms was eminently testable and falsifiable. Rather, it was evolutionary theory that could not be falsified, something that Popper himself had initially believed. There was, however, little or no attempt on any side of the debate to export notions of testability or falsifiability or scientific method more generally into the ordinary world of most Americans. Such notions at no point became bridges to involve ordinary people in science or to export the intellectual power of science beyond its immediate confines. On the contrary, Fred Spilhaus, executive director of the American Geophysical Union, claimed in 2005 that scientists and students were "bound by the scientific method" inside "their laboratories and science classrooms," but once outside of these special locations they could "believe what they choose about the origins of life."[20]

In some cases, even talking about methodology outside of the controlled environments of labs and classrooms could be dangerous. While supporters and critics of ID expressed general consensus on the importance of testability broadly conceived, there was often tension around the concept of "theory," just as there had been during the 1920s and 1930s. This tension was not directly related to the potential of violating boundaries between science and other kinds of knowledge, whether art or philosophy, but rather to the gap between scientific and popular usage. Those who worked to defend Darwin frequently complained that their opponents called evolution "just a theory," substituting a false sense of uncertainty for what was really reliable and secure. Such often intentional misinterpretation arose, according to those who decried it, because the word "theory" had "such a different meaning in a scientific context" compared with a colloquial one; its continued misuse demonstrated that "many people either do not know or do not accept the scientific definition" of the term.[21] Announcements that stressed the theoretical and thus uncertain status of evolution did occur, often at the grass-roots level of the ID movement, and typically in response to assertions, like the one offered by philosopher Michael Ruse, that Darwinian evolution was a "fact, *fact*, FACT."[22] In contrast, many of the leading figures and supporters of ID who aspired to a measure of scientific respectability used theory in a neutral or positive sense, regularly referring to ID "theory" and to Behe, Dembski, and others as ID "theorists." Nevertheless, the potential for misinterpretation appeared so severe to Lawrence Krauss, a physicist from Case Western Reserve University, that he suggested scientists should abandon using the term entirely in public discussion.[23]

Disagreement over what was typically called "methodological naturalism" was even more intense than differences over the nature of theories and facts. Evolutionists and their supporters often emphasized the restriction of legitimate science to purely natural explanations of the physical world, sometimes equating that restriction with the "scientific method."[24] At other times, natural explanations of the world were privileged because they were presented as uniquely testable, unlike beliefs based on faith in the supernatural. Sue Gamble, one of the Kansas State School Board members who opposed revised science standards that dropped reference to naturalism, bluntly asserted that "once you have supernatural explanations, you no longer have science."[25] For proponents of ID, such as Michael Behe, however, methodological naturalism was "nothing more than an artificial restriction on thought."[26] Others noted that "the attempt to equate science with materialism is quite a recent development, coming chiefly to the fore in the 20th century."[27] Instead, as was true in Kansas, ID enthusiasts stressed a wide range of other methodological activities. Gene Edward Veith, Jr., culture editor of the Christian-oriented *World Magazine*, stressed that true science "pursued the evidence wherever it leads." Johnson similarly called those "willing to consider evidence for ID" the "true empiricists and hence the true practitioners of scientific thinking."[28]

Methodological naturalism was particularly valuable, according to many of its boosters, because it helped to draw the line between the naturalistic realm of science and the supernatural world of religion. Indeed, by the turn of the twenty-first century the distinction between science and religion was largely unquestionable. Of course, numerous Americans followed their forebears in taking the opportunity of this distinction to show how scientific and religious knowledge harmonized or even worked in concert to increase human understanding. A publicly distributed e-mail message sent by the chancellor of the University of Kansas rejected attempts to dislodge evolution from public schools as "anti-science" but also stressed that there was no contradiction between evolutionary accounts of life "and a belief in God."[29] Likewise, a 2005 exhibit on Darwin at the Museum of Natural History, reviewed in *New York Magazine* and the *New Yorker*, included "a video of seven naturalists speaking about how they reconcile science with faith."[30] In a number of instances, Einstein took center stage as a symbol of the ways in which "the profounder sort of scientific minds" almost always contained a "religious feeling." Einstein provided a particularly congenial source of quotations for advocates of ID. Sympathetic authors sometimes recounted the German physicist's comments about his own "rapturous amazement at the harmony of natural law, which reveals an intelligence of such superiority that, compared with it, all the systematic thinking and acting of human beings is an utterly insignificant reflection."[31]

By and large, however, such attempts at harmonization were overwhelmed by a refrain of affirmations that science and religion were "best kept in separate realms," operated in "separate spheres" by whatever measure, or that one was the "opposite" of the other.[32] In an article for the liberal magazine the *Progressive*, political satirist Will Durst asserted that the Bible "has as much to do with science as gummy bears have to do with aerospace navigation. As far as science goes: It makes a great paper weight."[33] In this atmosphere, the religious pedigree of ID became a major preoccupation in determining whether it was "religion masquerading as science" or not.[34] One observer claimed that "an honest person" would recognize ID for what it was: "theology, not science."[35] The oft-cited separation of church and state only enhanced the potential power of such attestations. As reported in news sources from the *Skeptical Inquirer* to the *Christian Century*, the judge in the Dover, Pennsylvania, trial ultimately supported attempts to bar ID from classrooms because it was "grounded in religion" rather than science.[36] Critics' insistence on calling ID creationism or, more creatively, "stealth creationism," "neo-creationism," or "creationism with a container load of aluminum siding tacked on" drew on similar findings in court cases during the 1980s that declared creation science was fundamentally religious.[37] Again, rather than contest the separation of the science and religion, many supporters of ID responded by arguing that their ideas were not religious, were not "based on the Bible or any other scripture," and had no relation, rhetorical or otherwise, with creationism.[38] Instead, it was evolution that was "a religious theory that has penetrated natural science."[39]

At times, the presumed differences between science and religion grew so substantial that they became unbridgeable sources of conflict rather than opportunities for harmonious cooperation. Though a longtime defender of evolution, Michael Ruse claimed that "science, or at least its leading spokespeople, tends to be strongly antireligious" and that "no subgroup of scientists is more vocal than biologists, including the evolutionists."[40] Indeed, authors defending ID frequently cited a number of provocative and seemingly hostile comments by outspoken evolutionists, especially Richard Dawkins' observation that Darwin "made it possible to be an intellectually fulfilled atheist" and philosopher Daniel Dennett's insistence that Darwinism was a "universal acid" that "eats through just about every traditional concept." ID enthusiasts, who generally had shared commitments to traditional religion and to science, did not find evidence of a global struggle between science and religion in these pithy quotations; rather, they showed how an overly aggressive materialism had come to infect modern science. But others expressed a more general sense of conflict. According to journalist Matt Taibbi, the rise of ID itself was "a deliberate satirical echo of the great liberal lie of the modern age: the idea that progressive science and religion can coexist."[41] Even Ruse, though he did not endorse antireligious rhetoric, did wonder if a close and honest look at the nature and limits of science and religion might "lead to an ever greater divide between them." Philosopher Niall Shanks openly doubted that the "problem of reconciling science and religion" had any "universally acceptable rational solution."[42]

Nor was religion the only example of not-science that partisans cited amidst debate over ID. A semicoherent category of pseudo-science was alive and well. On the occasion of comments by President George W. Bush that seemed favorable toward including ID in public schools, journalist and author

John Derbyshire blasted the "teaching of pseudoscience in science classes" and asked, "Why not teach the little ones astrology? Lysenkoism? . . . Forteanism? Velikovskianism? . . . Secrets of the Great Pyramid? ESP and psychokinesis? Atlantis and Lemuria? The hollow-earth theory?"[43] Gary Wheeler, executive director of the National Science Teachers Association, also charged Bush with distracting Americans with a "pseudoscience issue."[44] Those same presidential remarks added impetus to preexisting claims that ID was "politicizing science," and thus "perverted and redefined" the pure nature of scientific knowledge.[45] Advocates of ID worked in some similar ways to show how evolution was something other than science as well. A 2005 article in *Newsweek* reported a claim made by the Discovery Institute's John West that "to say, as Darwinians do, that everything has to be reduced to a chemical reaction is more ideology than science." Similarly, an author in the conservative *American Spectator* assigned the claims of Darwinian evolution about the nature of humankind to the "realm of philosophy, not science."[46]

Discussion of ID was especially shot through with depictions of one the most yawning of all rhetorical chasms between science and what it was not: the one between scientific and popular worlds. Supporters of evolution sometimes examined the results of public polls that showed relatively low levels of belief in Darwinian evolution and decried the "enormous gap between the firmness of scientific evidence for evolution and the American public's beliefs about it" or lamented the "failure of scientists and science educators to convey the nature of science and its nature to the American public."[47] Journalist Chris Mooney and professor of communication Matthew Nisbet located some portion of that failure in the practices of their fellow journalists who had not specialized in reporting on scientific topics and thus used standards and styles, such at the tendency to strive for balanced views, that were not appropriate in talking about science.[48] There was somewhat more ambiguity among supporters of ID about the distinction between scientific and popular realms, though their focus on institutional affiliation and formal credentials did tend to implicitly inscribe one. And though both Behe and Dembski self-consciously wrote books for general audiences, they marked off segments of their work that they warned were intended for those with technical backgrounds and could safely be passed over by ordinary readers.[49]

This gap between scientific and popular realms was often one of meaning. In order to bolster his own prescription that scientists avoid "theory" because of differing definitions in scientific and popular realms, Lawrence Krauss cited the similar suggestion of Eugenie Scott that her fellow scientists and educators "should train ourselves to not use the term 'believe' in a scientific context because it blurs the distinction between science and religion."[50] The popular and scientific were distinguished not only by the ways people talked but also by how they thought. Not only did members of the general public potentially misinterpret reference to belief, they did not understand that "belief takes no part in scientific thinking."[51] To make matters worse, a number of observers reminded readers that science was filled with "counterintuitive" ideas and sought to overcome "naïve impressions" rather than confirm common sense.[52] Finally, descriptions of the structure of the popular realm emphasized its unfriendliness to scientific knowledge or practice. According to Barbara Forrest and Paul Gross, the movement that coalesced around ID was "*one of the most remarkable examples in our time of naked public relations management substituting successfully for the knowledge and facts of the case—substituting for the truth.*" Elsewhere they noted that among attempts to influence public policy or cultural patterns in the "public relations–driven and mass-communications world," "serious inquiry and questions of truth are often a mere diversion."[53]

Transgression and Transcendence

The last two hundred years of science talk have given Americans a wide array of rhetorical tools for distinguishing science from what it was not, and those engaged in public discussion of ID put them to work again and again. A science so distinct could sometimes seem unlike any other activity or branch of knowledge and, as was true elsewhere, this was one of its potential strengths. One observer of debate over ID claimed that "science will survive the challenge of intelligent design because science is the only branch of human endeavor that does not rely on humanity for its existence. It relies on the world outside of humanity, which is why it has been so successful."[54] The broadening of contrast between science and other merely human realms of thought beyond the religious sometimes carried with it the well-worn metaphor of warfare. The so-called "science wars" of the 1990s, during which self-identified defenders of science lashed out against scholarly attempts to account for the generation of scientific knowledge in terms of supposedly extrascientific factors such as politics or gender, reflected a tendency to link the power of science with its transcendence of ordinary human endeavor. Many advocates of ID similarly sought the sanction of the scientific because it had made other areas of thought "at best soft, sentimental stuff that may have a place in the church or synagogue but not in a serious domain like science."[55] Nancy Pearcey, an author and senior fellow at the Discovery Institute, complained that the separation of "science and religion into hermetically sealed compartments" had reduced the latter to a "noncognitive" category, "unconnected to questions of truth or evidence."

ID was so important, she continued, because it helped remove Christianity from "the sphere of noncognitive values" and restore it to the "realm of objective facts," ultimately allowing Christians to "effectively engage in the 'cognitive war' that is at the root of today's culture war."[56]

But, as they had over the last century and a half, the kinds of science constructed with the commonly available tools of science talk could also make it difficult to forge connections between the scientific world and the rest of American culture, Pearcey's hopes notwithstanding. Supporters of evolution who depicted themselves as defending science almost never justified their labors on the basis of the broad intellectual or ethical relevance of scientific ideas. Rather, on numerous occasions, they cited a presumed connection between teaching Darwinian evolution, often as a proxy for teaching science more generally, and the technological fruits of scientific research. Zoologist and educator Robert George Sprackland asserted that "America will continue to fall behind in medicine, technology, and other fields of science as long as our children are denied a good science education—one based on an understanding of what is, and is not, science."[57] In an article on how evolutionists could frame their criticisms of creationism in productive ways, David Morrison, a planetary scientist at NASA's Astrobiology Institute, imagined similarly appealing to an audience of parents by linking the teaching of evolution to American success at international competition in science and engineering. "We can make a persuasive and nonpartisan case," Morrison wrote, "that the future of the nation depends on its scientific and technical literacy."[58] Opponents of evolution also sometimes saw technology as one of the least problematic means of making science broadly important and relevant. An early draft of the revised Kansas science standards offered by one member in collaboration with a local creation science organization focused on teaching "technological science" and avoided "theoretical science," including Darwinism, as speculative and unreliable.[59]

In fact, outside of reference to technological applications, the very same turn-of-the-twenty-first-century science talk that made science a valuable category could also place serious limitation on the scientific and confine its potential influence. Again, this was the culmination of a trend visible since the late 1800s. Not that all restrictions were bad. Stressing the limitations of science provided one useful means of defending it from charges of hostility toward religion or cultural imperialism more generally. Eric Cornell, a physicist and Nobel laureate, called on scientists to resist the inclusion of ID in science classes. At the same time, he urged his colleagues "not to say that the God answer" to questions about the natural world "is unscientific," but rather "that the methods of science don't speak to that answer."[60] Philosopher John Boler made a similar point in an article for the Catholic magazine *Commonweal*, advocating the exclusion of ID from science classrooms while admitting that "many important areas of discussion fall outside the purview of science; most issues about values are not scientific."[61] In discussing his own thoughts on the debate over ID, Michael Ruse stressed that "those who love science (and I am one), including those of us who are nonbelievers (and I am one), should quit sneeringly giving religion the backs of our hands and start to look seriously at the limits of science."[62]

But, taken from a slightly different angle, these very same claims could show how science might not be relevant to many or even the most important issues in the modern world. In a short article about science and faith in *Time* magazine, Francis Collins, director of the National Human Genome Research Institute, claimed that "science's tools will never prove or disprove God's existence." He went on to admit that "for me the fundamental answers about the meaning of life come not from science but from a consideration of the origins of our uniquely human sense of right and wrong, and from the historical record of Christ's life on earth."[63] There were even times when a limited view of science severed it from the larger world of truth of which it had once been a seamless part. In the course of his critique of ID as religious and therefore unscientific, Robert George Sprackland capped his own definition of science by declaring that "science is not about the search for truth or proving the existence of anything; rather it is a process of interpreting and understanding nature without recourse to the supernatural."[64] Such a claim would have been nearly unthinkable two centuries earlier and, taken on its own, made the threat of being unscientific less of a major blow and more of a minor inconvenience, perhaps even a badge of honor.

Talk about the limits of science was particularly strong in the rhetoric of many advocates of ID. Like an earlier generation of ufologists, those friendly to ID often had a very complicated relationship with science. On the one hand, they portrayed themselves as making scientific claims about the world. On the other, they frequently felt compelled to explain the rejection of those claims by the bulk of the orthodox scientific community. In *Darwin's Black Box*, Behe admitted that a poll of scientists would reveal that the vast majority believed in Darwinian evolution. He claimed, however, that most of those believers did so "based on authority" and that, furthermore, "too often criticisms have been dismissed by the scientific community for fear of giving ammunition to creationists."[65] A variety of others followed this same line of attack even more intently, depicting the exclusivity of the "scientific establishment," a term widely used in ID literature, as a source of unreliability. Tom Bethell, senior editor of the conservative *American Spectator*, scorned the view that a "white-coated priesthood" of scientists should be heeded on all matters "because they have the measuring devices." Rather, he offered a depiction of a scientific community "more interested in their own funding, tenure, and security than in any detailed accounting of progress or decline in their own field."[66]

At other times, advocates of ID called on work in the history and sociology of science, or even the reflections of practicing researchers, to argue that scientists as a group were generally so concerned with maintaining their social position that they could not acknowledge obvious truths. In a 2004 collection of essays edited by William Dembski, lawyer Edward Sisson pointed to the work of sociologist Bernard Barber "in which he discussed the fact that, on occasion, the scientific establishment may resist evidence that challenges a prevailing theory."[67] Dembski himself echoed a similar sentiment when he reflected on his experiences with the system of peer review. Those experiences, he wrote, confirmed an observation made by none other than Paul Gross: "Being right isn't enough. What you say, however right, must be said in currently acceptable language, must not violate too brutally current taste, and must somehow signal your membership in a respectable professional club."[68] Others pointed to the postwar work of historian and philosopher Thomas Kuhn, who suggested that scientific work was guided by ruling "paradigms" that shape (or alternately limit) the sorts of facts and theories that practitioners could bring themselves to accept.

While the strict boundaries around the scientific community could become useful tools for explaining away resistance, ID itself as an intellectual, social, and cultural movement was even more deeply invested in the limitations of scientific knowledge. Unlike scientific creationists, who sought to use scientific tools to reveal divine power in action, the leading figures of ID founded their ideas on the admission that science could not go so far. Dembski noted without much hand wringing that "design theorists recognize that the nature, moral character and purposes of" the intelligent designer "lie beyond the remit of science."[69] The happy acceptance of this limitation supposedly distinguished ID from creation science, and religion more generally, thus providing one basis of its scientific pedigree. But advocates of ID, unlike so many defenders of evolution, did not rest with the casual acknowledgement that science could not lead to God; rather, they sought to fix the borders of legitimate science, to establish them once and for all, so that though purely scientific explanations could not speak to the existence of something beyond nature, their absence might. Behe not only claimed that "molecular machines," such as the bacterial flagellum, "raise an as-yet impenetrable barrier to Darwinism's universal reach," but also that "that complexity of life's foundations has paralyzed science's attempt to account for it."[70] Locating the divine in nature through such means resembled what has often been referred to as the God-of-the-gaps approach, one in which God's action was associated with those phenomena that science could not as yet explain. ID, however, did not simply accept gaps as given or as vulnerable to a new discovery; it was fundamentally about creating and establishing such gaps in Behe's irreducible complexity or Dembski's mathematical tool for detecting intelligent engineering, thus ensuring that they would not close.

In light of such a confined view of science, supporters of ID found themselves straying again and again from purely scientific ground, as they depicted it, to talk about the issues that were fundamentally important to them. Both Dembski and Behe included discussions or sections in their books that explicitly dealt with what they presented as larger philosophical and religious questions.[71] Specific issues regarding the basic intelligibility of the world, with which ID grappled, were "as much moral and aesthetic as it is scientific," according to Dembski.[72] Even more fundamentally, many of ID's most vocal enthusiasts made it clear that there was "much more at stake . . . than a scientific theory" in their battle against evolution. Nancy Pearcey attributed popular interest in Darwinism and its limitations to the public's innate sense that such matters were "really about much more than science" and dealt with issues "far beyond science."[73] The materialism lurking in purely Darwinian evolution threatened "potentially grave moral and political consequences" by, for instance, providing the justification for a range of behaviors from "abortions for sex selection, infanticide for children with disabilities, and the growing practice of assisted suicide." After all, argued John G. West, Jr., assistant director of the Center for Science and Culture, "if humans are purely material beings, then their worth inevitably degrades as their physical capacities deteriorate."[74] Ultimately, claimed Dembski, "[n]aturalism promises to free humanity from the weight of sin by dissolving the very concept of sin. . . . Yes, we pollute the earth and decimate rain forests and cause plants and animals to go extinct. But all of this is in accord with nature's laws, not in violation of them."[75]

These effects were one basis on which evolution was frequently criticized as philosophy, ideology, or even religion. But at the same time, believers hoped that ID would do more than simply replace evolution in the realm of science. Rather, many of them expressed the hope that its benefits would "spread far beyond the domain of science" and would "create a potent force for cultural renewal in the twenty-first century."[76] The initial name of the Center for Science and Culture—the Center for the Renewal of Science and Culture—made such aims explicit. Yet an ID charged with such a mission often became itself something other, or rather more than scientific. Lawyer Dan Peterson claimed in the *American Spectator* that the contest between evolution and ID was ultimately one of "two competing philosophical visions."[77] Likewise, Dembski noted that "we are dealing here with something more than a straightforward determination of scientific facts or confirmation of scientific theories. Rather we are dealing with competing worldviews and incompatible metaphysical systems."[78] There were even times when ID took on the trappings of religion without toppling into the dustbin of not-science. David Klinghoffer, an author and senior fellow of the Discovery Institute, asserted that the religious backgrounds of many of ID's supporters did not delegitimize the theory any more than the materialistic beliefs of some evolutionists discredited Darwin; rather, both were "expressions of a certain kind of faith."[79]

Such rhetoric, grounded in the boundedness of modern science talk, made science simultaneously something to be sought after and something to be transcended. This ambivalence could be and was frequently ignored. But a few advocates of ID resolved the tension between these depictions of science in the same way that many believers in UFOs did several decades before. In short, they turned to a self-consciously alternative vision of the scientific, one that was large enough to contain their ideas and aspirations. Behe, perhaps because of his continued work within and appeal to the structures of orthodox science, did not, or at least has not as yet, depicted ID as doing anything more than reinvigorating "a field of science grown stale from lack of viable solutions to dead-end problems."[80] Dembski, however, who might be cast as Jacque Vallee to Behe's J. Allen Hynek, was much more explicit about ID's larger battle against an "inadequate conception" of the nature of the scientific and its promise for creating a "science of the new millennium," pursued by a "new generation of scholars and professionals who explicitly reject naturalism and consciously seek to understand the design God has placed in the world." Indeed, he rejected the sufficiency of a more piecemeal approach—for instance, critiquing the tendency in established science to reduce everything to purely material mechanisms—because such efforts would do "nothing to change science—and it is science that must change."[81] Others looked forward to the possibility of an ID-inspired paradigm shift in science, that is, a fundamental alteration in the framework of basic assumptions and ideals that scientists use in their work, or to the potential of ID to "revolutionize science as dramatically as Newtonian physics did in the first scientific revolution."[82]

The rhetoric of ID's supporters interested in remaking the scientific enterprise also possessed some similarities to the styles of talk used by ufologists and others seeking to create an alternative science. Both were sometimes self-consciously transgressive, not only according to critics but also to advocates. John Wilson, the editor at large for *Christianity Today*, warned readers of a collection of pro-ID essays that "by questioning Darwinism, you place yourself in the company of all the cranks who have violated taboos enforced by our current opinion-makers."[83] Depictions of battle against an entrenched scientific or Darwinian "establishment" and claims that ID theorists pursued "alternative scientific theories and research methods" conveyed similar, though somewhat more muted, impressions. The alternative science that occasionally coalesced around ID, again like the one that frequently formed around UFOs, was also far more inclusive than its orthodox reflection in ways that reclaimed ground lost during the previous two centuries. For instance, Dembski portrayed theology, which he noted was "traditionally called 'the queen of the sciences,'" as the one branch of knowledge that "transcends, informs and unifies" all the others and claimed that "all disciplines find their completion in Christ and cannot be properly understood apart from Christ."[84] There was also a note of populism in some depictions of the new science represented by ID. Pearcey praised ID as a "concept that is simple, easy to explain, and based solidly on experience."[85] Others pointed to its intuitive correctness.[86] Finally, a science remade by ID was both capable of revealing transcendent truths and also relevant, answering "the public's most pressing concerns" in ways that orthodox science was not able to. A correspondent in the *Texas Monthly* asked, "What good does it do to raise our C in science to an A when we learn things that can't be proven? Intelligent design can't be proven either, but at least it gives us a philosophy that can be lived out in everyday life."[87]

There was also a major difference between depictions of the new sciences engendered by UFOs and ID. In the latter case, even though the aim was far above the material universe, much of its rhetorical foundation was planted squarely in the realm of technology. Certainly, many believers imagined flying saucers as technological artifacts, but UFOs often became part of a full-fledged vision of an alternative science as paranormal, even supernatural objects rather than nuts-and-bolts examples of extraterrestrial equipment. In the case of ID, technological metaphors were, while not always immediately apparent, central. Behe based his ideas fundamentally on the notion of biochemical structures as "molecular machines" and called attention to the recent artificial design of biochemical systems by human engineers as a precedent for his own interpretations.[88] As he reflected on the "new kind of science" that ID would enable, Dembski noted that "the reason evolutionary biology has lost all sense of proportion about how much evolution is possible as a result of blind material mechanisms (like random variation and natural selection) is that it floats free of the science of engineering." Instead, ID sought to understand biological organisms as "engineering systems." Elsewhere he highlighted one of ID's positive contributions to science as allowing researchers to "reverse engineer objects shown to be designed" and stated that "the design theorist is a reverse engineer."[89] Nor were Behe and Dembski alone in linking ID to engineering. A letter to *Electronic Engineering Times* asserted, "Naturalistic evolution is the antithesis to engineering. Engineers understand that complex structures are intelligently designed, not the product of random variations."[90] Such images suggest the intensification of technological rhetoric in the general repertoire of American science talk, even in the realm of the unorthodox.

Creating Science

By the early years of the twenty-first century, science was thus a term with many possible meanings, ones that partisans debating ID were all too willing to put to work. The flexibility of science was nowhere more evident than in claims about the general status of scientific knowledge and methods. Not surprisingly, given their desire to place non-naturalistic biological design on a scientific footing, many advocates of ID stressed the "tremendous cultural prestige" that science held in the modern "age of technology."[91] Dembski thought it bore repeating that "the only universally valid form of knowledge within our culture is science" and that "neither religion, philosophy, literature, music, [nor] art makes any such cognitive claim."[92] At times, this perceived cultural authority could threaten to go too far. Edward Sisson suspected that high-school students encountered evolution without a discussion of its weaknesses "so that deference to scientists becomes a foundational assumption in their mental makeup *before* they are old enough to question either the theory or the authority of the scientific establishment that is built on it."[93] Meanwhile, some defenders of Darwin reached the opposite conclusion. In *Creationism's Trojan Horse*, Barbara Forrest and Paul Gross complained that "in the west, opinions, perceptions, loyalties, and, ultimately, votes are what matter when the goal is to change public policy—or for that matter, cultural patterns." The search for truth was often beside the point.[94] In other cases, observers linked the rise of ID, criticism by the religious right and the postmodern left, and abysmal public knowledge of the basics of evolution with the sinking status of science in the United States. A columnist in *Esquire*, for instance, juxtaposed increasing support for ID with what he called the rise of "Idiot America" and its "war on expertise."[95]

These conflicting assessments bring us back to the "paradoxical" place of science in American culture. If determining the status of science were simply a matter of measuring the amount of money the federal government spent to support organized and original research, counting the number of scientists consulted on matters of public policy, or plumbing the depths of popular knowledge about particular topics, we might clear up any confusion with ease. If the nature of the scientific were as accessible as the nearest dictionary, the content of public school curricula, or the judgment of professional practitioners, serious scholars, or well-meaning judges, we could map the limits of its influence without undue toil. Science is of course more, or more accurately perhaps something other than any of these things. We can say with some confidence that whatever meaning Americans have attributed to science in the modern world, they have tended to do so with rhetorical tools that made it appear either uniquely powerful or specially handicapped because it was so unlike any other body of knowledge. The distinction between differing evaluations of the cultural authority of science among people arguing over ID came down ultimately to which scientific boundaries they wanted to stress and for which purposes: the ones that divided science from and typically elevated it above religion or the ones marked by the existence of a methodology, a body of knowledge, or a set of values of which the general public was ignorant almost by definition. The former notion of the scientific provided solid ground on which to establish ideas that would otherwise be mired in shadowy and subjective realms. The latter could challenge popularizers to redouble their attempts at reasserting the broad value of science.

Accepting the view that the meaning of science is actively constructed, not only by scientists but by everyone, certainly carries risks, just as the acceptance of democracy does in the political sphere or unrestricted free markets do in the economic one. There are a variety of good reasons for thinking that a harmonious, fair, and compassionate civilization demands various limitations and restrictions. Depriving the scientific of a stable and universal referent leaves a central conceptual category vulnerable to the machinations of those with sufficient social power, skills at public relations, and inevitably a variety of unsavory social agendas. It also raises the specter of relativism, in which individuals are isolated in their own personal worlds, though asserting the essential meaning of key terms hardly does more to bridge divisions or promote dialogue. Rather, it frequently justifies the reverse. It cannot be surprising that a few partisans on both sides of debate over ID expressed discomfort that words might be able to float free of their supposedly established meanings. Political scientist Ron Carstens objected to proposed changes in school standards' definitions of science by warning that "if the language doesn't stay universal, science can mean anything you want it to mean."[96] Dembski was especially critical of the "deconstructionist claim that 'texts are indeterminable and inevitably yield multiple, irreducibly diverse interpretations.'" Such a view smacked of pragmatism, which in turn drew its strength from naturalism's elimination of a "transcendent realm of meaning to which our linguistic entities are capable of attaching." Rather, for Christian theism as Dembski imagined it, meaning "transcends convention" and language is "a divine gift for helping us to understand the world and by understanding the world to understand God himself."[97]

A vision of information flow that sacrifices flexibility and freedom in favor of the ability to pronounce on what was happening in other people's minds, often as a prelude to changing them, has proved particularly attractive to a range of actual or aspiring intellectual elites, scientific, political, commercial, scholarly, and otherwise. Despite their occasional populist rhetoric, their desire to offer American schoolchildren an alternative to evolution, and the Discovery Institute's espousal of free-market principles, advocates of ID were no more above taking a highly restricted and essentialist view of meaning than defenders of orthodox evolution and "real" science. Nancy Pearcey, for instance, took aim at the concept of Dawkins' memes as a key component of evolutionary psychology. And yet she and others described evolutionary notions in almost memetic terms that dismissed the possibility that anyone could freely choose to believe them. Pearcey asserted that evolution formed part of a "total worldview" and echoed the claim of another supporter of ID that "a naturalistic definition of *science* has the effect of indoctrinating students into a naturalistic *worldview*."[98] In a number of other works, Dembski depicted naturalism as a "toxin that pervades the air we breathe and an infection that has worked into our bones" and portrayed the aggressive promotion of Darwinism in classrooms as an attempt at "total worldview reprogramming."[99] Behe likewise identified the "inculcation of an evolutionary worldview" in modern teaching of biology. Thankfully, according to Johnson, ID offered a nearly certain promise of challenging the "cultural power of evolutionary naturalism" and the way it "dominates the contemporary mind."[100] It was not always clear how ID would achieve such a feat without some inculcation or reprogramming of its own, though given its righteous ends such means might have been excusable.

By promoting a flexible and negotiated view of meaning, I am certainly not suggesting that public commentators give up talking about the nature of science. As is the case with any important keyword, from "marriage" to "free speech," defining science is an especially important and vital activity in the context of institutions with the power to enforce rhetorical divisions between different kinds of things. Centuries of boundary-work, including recent struggles over ID, have made clear that the accepted nature of science encoded into the rules and regulations of the public schools, courts, or federal funding agencies has profound implications for the structure of, the flow of resources through, and the balance of power in society. But, so far, no such strategy has done much to enhance the quality of popular discussion of scientific issues more generally, nor is it likely to do so. Once beyond the reach of institutions where particular visions of the scientific can be imposed, Americans, including scientists, espouse whatever makes the most sense to them. In a highly bounded culture, audiences will normally have the further means to subvert or reject what does not. Reflecting on his many years of teaching, physicist Lawrence Lerner acknowledged that many students "come to believe that the real world and the 'physics-class world' operate according to different laws." They learn enough to pass tests, but "feel no need to reconcile" what they learn about that world in school with "the one in which they drive their cars and generally live their lives."[101] The same is almost certainly true for what people learn to say about science while working in different institutional contexts.

Rather, accepting talk about science as an ongoing and semiautonomous cultural enterprise suggests that there are definite limits on what scientists, popularizers, and educators can expect to gain by trying to "educate the public about the nature of science itself," whatever that might be, even less by demanding that people adjust to a particular vision of the scientific.[102] As Ralph Waldo Emerson once wrote, "[L]anguage is a city to the building of which every human being brought a stone."[103] Those who claim to speak for scientific knowledge in public realms cannot claim to speak for science as a cultural category in quite the same way. They participate in its construction rather than totally determine it. Science as a category exists because we construct and reconstruct it thousands of times a day. Such efforts are not mere word games, but rather means of creating, within the limits of the rhetorical tools available and the idiosyncratic ability to make new ones, the world we live in. Particularly in those controversial areas where significant social and cultural issues are often at stake, the communal and negotiated nature of science means that the ground of scientists' public expertise is partly established by people other than scientists themselves. Of course, the good news is that the meaning of science will continue to be negotiated in the general marketplace of ideas, and that admirers and defenders of science are as free to participate in that negotiation as actively as anyone else.

One of the most active areas of change in present-day science talk appears to be the relationship between science and technology. So significant has the rhetoric of technology become to making science relevant and important that the technological has sometimes eclipsed the scientific, just as the scientific once began to obscure the harmonious world of truth that formerly stood behind all specific forms of human knowledge. As I have already noted, many Americans began to engage in a kind of linguistic expansion of science in the decades straddling the turn of the previous century. References to scientific medicine, scientific engineering, scientific management, scientific advertising, and even scientific motherhood all spread. Today, appeals to scientific "anything" are somewhat more rare. Instead, we find talk of biotechnology, information technology, nanotechnology, and even someday soon, perhaps, cognitive technology. Among modern intellectuals especially, technological metaphors abound, whether allusions to systems, platforms, construction, or "technologies" as general methods of working. "Technoscience" has already achieved widespread use by sociologists of science to refer to the intertwined production of abstract knowledge and material devices. We will see whether a century or two from now, coming generations will replace reference to science entirely with a term that strikes them as more relevant, powerful, and meaningful, whether technoscience or something more fanciful and exotic, like "techknowledgy."

Alternately, some public commentators have sought to more consciously increase the prominence of science in the lives of ordinary people. In some ways, these attempts have involved something like a step backward in time to an age in which the links between science and other realms of thought were far more pervasive. Recent European concern over a public backlash against nanotechnology in the mold of widespread rejection of genetically modified foods has spread proposals for a new kind of "see-through" science that is not separate from public concerns but incorporates them at the earliest stages of research and development, possibly though such democratizing mechanisms as citizen juries.[104] The ultimate success of what amounts to an attempt to return science, at least in some measure, to the status of a communal possession remains to be seen. It has arisen in a field with stronger rhetorical links to technology, where market forces and public participation are far more commonly accepted, than to science. Reworking of the relationship between science and the public can also call into question the very separateness that has made science an important keyword in the first place. Indeed, the debate over scientific knowledge as opposed to applications sparked by ID has not generated any serious suggestions of citizen juries or calls for see-through science and has generally sparked reflection on the proper limits of democracy rather than its extension.

Scientific representatives who want to speak about broad social and cultural issues may find their greatest success part way along a path already blazed by advocates of alternative science; they may need to speak for something more than science, whether that means their personal convictions and concerns or the social, political, or religious movements to which those convictions and concerns have led them. Accepting the limits of scientific boundaries, rather than trying to breach them, may ultimately encourage a whole range of new occupations along the lines of science journalism or the history of science that combine scientific training with experience and commitments in politics, advertising, religion, art, or other disciplines. Stepping away from purely scientific ground may make scientists seem less aloof from the social and political divisions that occur around them, and may seem to jeopardize the ability of science to provide neutral and objective truths. But, if approached openly, honestly, and with humility, it could also become the basis, to paraphrase Burnham, for reanimating Einstein—that is, reincorporating American scientists, not so much as scientists but as citizens, into broadly public discussion. Increasing intellectuals' comfort with engaging social, political, and religious divisions not as analysts but as earnest and open participants can perhaps also rescue them generally from their current oblivion in American culture.

One more place where popularizers and educators might do well to step back from science is in the promotion of a uniquely scientific method. Certainly, as the last two centuries have shown, such a concept is a valuable and useful one and I do not expect that Americans will stop invoking it. But it has also come with a cost. In many ways, a passive and frequently irrational popular mind has roots in strongly bounded depictions of scientific thinking that drove the capacity to make rational judgments about the nature of reality beyond public reach. At the very least, the notion that there is some particular mode of thought that is the sole possession of scientific practice, that must be adopted to consider scientific questions, and that can only be learned through education in some scientific subject has narrowed the ground of the ordinary person's ability to engage science. Rather, there is no reason to believe that the basic critical thinking skills that can be distinguished from the discipline-specific and often highly technical practices within individual sciences are not precisely the same ones required to do history, analyze literature, or make sense of political issues. Indeed, it may be easier to learn such skills in a more accessible and immediately relevant area such as history rather than in science, where introductory material is often very far removed from the sorts of subjects that become the topics of serious public debate. To focus on critical thinking rather than scientific method, even in the judgment of scientific questions, is to slide back to a semi-Baconian view, not so much because of an emphasis on facts but because of the sense that the ability to think well transcends any single topic.

Decoupling critical thinking from scientific method is one way of helping to enable ordinary people to become capable and critical consumers of information, scientific and otherwise. Being a critical consumer of scientific information does not necessarily mean thinking like a scientist or even having a particularly good basic knowledge of science. It involves having access and the confidence to use tools to evaluate claims to expertise and the kinds of people who make them, decode the social and culture commitments behind the construction and use of various categories such as science, judge evidence offered to justify the decisions of others as well as their own, and draw sound and reasoned conclusions about the effects of various ideas on their own lives and the larger well-being of their communities. In many ways, such abilities are far more valuable in the modern world than understanding Newtonian mechanics or natural selection. Nonscientists are far more often confronted with science-wielding experts, or with people claiming to be experts, than with technical problems that they are asked to solve on their own. Well-prepared information consumers may not choose science, or at least the kinds of science that advocates of one kind or another might wish. But they would be much more capable of understanding the meanings and repercussions of the choices they do make, and would make whether well prepared or not.

Finally, none of these suggestions and reflections is intended to question the value of science popularization itself. There is considerable value in offering nonscientists opportunities to understand the contents of contemporary scientific knowledge or the general outlines of modern scientific practice. Such knowledge is in fact an essential contribution to the modern information landscape. Present-day public discussion would be poorer without it. Nevertheless, defenders of science should not encourage nonscientists to use the label of science as a substitute for a complete discussion of the accuracy and reliability of knowledge claims or the practices that produced them. They should certainly not provide an example of such usage themselves or confuse science as a cultural category with the empirical knowledge about the physical that is typically associated with it. Taking a step back from claims about the objective existence of something called science can ultimately help to promote greater dialogue by helping us to pay attention to the ways in which we ourselves construct categories, to examine the claims of opponents in the same spirit as our own, and to avoid dismissively attributing the views of others to limiting worldviews that we are able to transcend. Accepting the often ambiguous nature of language is in that way a step toward encouraging us to actually listen to one another much more then we currently do.

Notes

Introduction: Talking about Science

1. Thomas Broman, "The Habermasian Public Sphere and 'Science in the Enlightenment,'" *History of Science* 36 (1998): 143.
2. Barry Barnes and David Edge, introduction to *Science in Context* (Cambridge, MA: MIT Press, 1982), 2.
3. George Daniels, *Science in American Society* (New York: Knopf, 1971), 293.
4. Frederick Lewis Allen, *Only Yesterday: An Informal History of the Nineteen-Twenties* (New York: Harper & Brothers, 1931), 197.
5. For a review of public opinion surveys on science, see Jon D. Miller, "Public Understanding of, and Attitudes toward, Scientific Research: What We Know and What We Need to Know," *Public Understanding of Science* 13 (2004): 273–94. For specific comments on positive evaluations of science, see ibid., 284–85.
6. Christopher Toumey, *Conjuring Science* (New Brunswick, NJ: Rutgers University Press, 1996), 6 and 153.
7. Marcel C. LaFollette, *Making Science Our Own* (Chicago: University of Chicago Press, 1990), vii.
8. Paul Kurtz, "The Growth of Antiscience," *Skeptical Inquirer*, Spring 1994, 255.
9. Ruth E. Malone, Elizabeth Boyd, and Lisa A. Bero, "Science in the News: Journalists' Construction of Passive Smoking as a Social Problem," *Social Studies of Science* 30 (2000): 716.
10. *Science and Engineering Indicators* (Washington, DC: National Science Board, 2002), 7.3.
11. Miller reports that during the 1980s and 1990s, between 10 and 17 percent of U.S. adults qualified as scientifically literate. See Miller, "Public Understanding," 288.
12. Richard Dawkins in *The Best American Science and Nature Writing 2003* (Boston: Houghton Mifflin, 2004), xvii. Italics in the original.
13. Jim Holt, "Madness about a Method," *New York Times Magazine*, 11 December 2005, 25.
14. Kurtz, "The Growth of Antiscience," 255.
15. Mike Michael, "Ignoring Science: Discourses of Ignorance in the Public Understanding of Science," in *Misunderstanding Science?* ed. Alan Irwin and Brian Wynne (Cambridge: Cambridge University Press, 1996), 116–17.
16. Ibid., 118.
17. Ibid., 120.
18. "Drop Science, Drop Knowledge," *Rap Dictionary*, 28 February 2005, http://www.rapdict.org/Drop_science%2C_drop_knowledge (28 March 2005).

19. Philip K. Dick, "How to Build a Universe That Doesn't Fall Apart Two Days Later," in *I Hope I Shall Arrive Soon*, ed. Mark Hurst and Paul Williams (Garden City, NY: Doubleday, 1985), 8.

20. For some basic works on the idea of professionalization, see George H. Daniels, "The Process of Professionalization in American Science: The Emergent Period, 1820–1860," *Isis* 58 (1967): 151–66; Nathan Reingold, "Definitions and Speculation: The Professionalization of Science in America in the Nineteenth Century," in *The Pursuit of Knowledge in the Early American Republic*, ed. Alexandra Oleson and Sanborn C. Brown (Baltimore, MD: Johns Hopkins University Press, 1976), 33–69; Sally Gregory Kohlstedt, "Savants and Professionals: The American Association for the Advancement of Science, 1848–1860," in *Pursuit of Knowledge in the Early Republic*, 299–325.

21. Historian Paul Starr defined cultural authority as "the construction of reality through definitions of fact and value." See Paul Starr, *The Social Transformation of American Medicine* (New York: Basic Books, 1982), 13.

22. On works that grapple with the contested nature of science, see Alison Winter, "The Construction of Orthodoxies and Heterodoxies in the Early Victorian Life Sciences," in *Victorian Science in Context*, 24–50; idem, "Mesmerism and Popular Culture in Early Victorian England," in *History of Science* 32 (1994): 317–43; idem, *Mesmerized* (Chicago: University of Chicago Press, 1998); James Secord, *Victorian Sensation: The Extraordinary Publication, Reception, and Secret Authorship of Vestiges of the Natural History of Creation* (Chicago: University of Chicago Press, 2000); Irwin and Wynne, *Misunderstanding Science*; Ronald L. Numbers, *The Creationists* (New York: Knopf, 1992).

23. *Towne v. Eisner*, 245 U.S. 418 (1918).

24. On the notion of keywords, see Raymond Williams, *Keywords: A Vocabulary of Culture and Society* (New York: Oxford University Press, 1976); idem, *Culture and Society, 1780–1950* (London: Chatto & Windus, 1969); Daniel T. Rodgers, *Contested Truths: Keywords in American Politics since Independence* (Cambridge, MA: Harvard University Press, 1998).

25. On boundary-work, see, Thomas F. Gieryn, *Cultural Boundaries of Science* (Chicago: University of Chicago Press, 1999); idem, "Boundary-Work and the Demarcation of Science from Non-Science: Strains and Interests in Professional Ideologies of Scientists," *American Sociological Review* 48 (1983): 781–95; Thomas F. Gieryn, George Bevins, and Stephen Zehr, "Professionalization of American Scientists: Public Science in the Creation/Evolution Trials," *American Sociological Review* 50 (1985): 392–409.

26. Gieryn, "Boundary-Work," 783–87.

27. See, for instance, Theodore Sasson, *Crime-Talk: How Citizens Construct a Social Problem* (Hawthorne, NY: Aldine de Gruyter, 1995); Martin Montgomery, "D-J Talk," in *Styles of Discourse*, ed. Nikolas Coupland (London: Croom Helm, 1988), 85–104.

28. On discourse in the media especially, see, for instance, Allan Bell and Peter Garrett, *Approaches to Media Discourse* (Oxford: Blackwell, 1998); John Hartley, *Understanding News* (New York: Methuen, 1982); Myra MacDonald, *Exploring Media Discourse* (London: Arnold, 2003); Teun A. van Dijk, ed., *Discourse and Communication: New Approaches to the Analysis of Mass Media Discourse and Communication* (New York: Walter de Gruyter, 1985); Norman Fairclough, *Media Discourse* (New York: Arnold, 1995).

29. Raymond Williams makes this general claim in "Base and Superstructure in Marxist Cultural Theory," in *Rethinking Popular Culture*, ed. Chandra Mukerji and Michael Schudson (Berkeley: University of California Press, 1991), 407–23.

Williams has also examined Anglo-American culture in terms of particular key-words. This notion that the popular culture is a reservoir of rhetorical tools is also consistent with an idea offered by David Hall that culture is composed of a repertory of languages. See Hall, introduction to *Understanding Popular Culture: Europe from the Middle Ages to the Nineteenth Century*, ed. Steven L. Kaplan (New York: Mouton, 1984), 14–16.

30. A focus on the rhetorical tools used to describe a given phenomenon rather than on specific descriptions is explored in Ann Swidler, "Culture in Action: Symbols and Strategies," *American Sociological Review* 51 (1986): 273–86. See also Ulf Hannerz, *Soulside: Inquiries into Ghetto Culture and Community* (New York: Columbia University Press, 1969).

31. Bert Hansen, "America's First Medical Breakthrough: How Popular Excitement about a French Rabies Cure in 1885 Raised New Expectations for Medical Progress," *American Historical Review* 103 (1998): 374–75.

32. LaFollette, *Making Science*, 2.

33. On Dawkins' original discussion of memes, see Richard Dawkins, *The Selfish Gene* (Oxford: Oxford University Press, 1976), 201–15. On more recent work on memes, see Richard Brodie, *Virus on the Mind: The New Science of the Meme* (Seattle, WA: Integral Press, 1996); Susan Blackmore, *The Meme Machine* (Oxford: Oxford University Press, 1999); *Darwinizing Culture: The Status of Memetics as a Science*, ed. Robert Aunger (Oxford: Oxford University Press, 2000); Robert Aunger, *The Electric Meme: A New Theory of How We Think* (New York: Free Press, 2002).

34. Roger Cooter and Stephen Pumphrey have pointed out not only the paucity of work on the influence of science on popular culture but also the equally important "and even more neglected . . . *lacks* of impact, and/or the struggles behind and resistances to popularization of science." See "Separate Spheres and Public Places: Reflections on the History of Science Popularization and Science in Popular Culture," *History of Science* 32 (1994): 338.

35. Muriel G. Cantor, review of *Making Science Our Own*, *American Journal of Sociology* 97 (1991): 260.

36. For an overview of work on the influence of media, see Justin Lewis, *Constructing Public Opinion* (New York: Columbia University Press, 2001), 77–102. On audience studies, see James Hay, Lawrence Greenberg, and Ellen Wartella, eds., *The Audience and Its Landscape* (Boulder, CO: Westview Press, 1996); Jon Cruz and Justin Lewis, eds., *Viewing, Reading, Listening* (Boulder, CO: Westview Press, 1994); Janice Radway, *Reading the Romance* (Chapel Hill: University of North Carolina Press, 1984); Sonia Livingston, *Making Sense of Television* (New York: Pergamon Press, 1990); Tamar Leibes and Elihu Katz, *The Export of Meaning* (New York: Oxford University Press, 1990); John Fiske, *Media Matters* (Minneapolis: University of Minnesota Press, 1994). During the nineteenth century, James Secord's reconstruction of the habits of British readers confronting the *Vestiges of the Natural History of Creation* (1844) showed reading as a very active and often synthetic process in which people often drew on ideas from many sources of information. See Secord, *Victorian Sensation*, 336–63.

37. Of course, different kinds of people had access to different sources of science talk, based on their personal preferences and their social and economic placement. Further studies can shed considerable light on the ways elements of science talk circulated differently in various communities and how those differences affected people's relationship with science. Rather than pursue such a course here, however, I aim to set a kind of baseline by examining those habits of talking that were most widespread and most likely to be familiar to the vast majority of literate Americans.

38. See, for instance, Noel Ignatiev, *How the Irish Became White* (New York: Routledge, 1995). On the invention of various significant categories in America, see Edmund Morgan, *Inventing the People* (New York: W. W. Norton, 1988); Helen Carr, *Inventing the American Primitive* (New York: New York University Press, 1996); Mary Beth Norton, *Founding Mothers & Fathers: Gendered Power and the Forming of American Society* (New York: Knopf, 1996).

39. On advertisers' interest in demographics and its impact on media content, see Gerald Baldasty, *The Commercialization of the News in the Nineteenth Century* (Madison: University of Wisconsin Press, 1992), 63–75. On the development of standard genres in American popular literature, see John G. Cawelti, *Adventure, Mystery, Romance: Formula Stories as Art in Popular Culture* (Chicago: University of Chicago Press, 1976).

40. On the development of advertising in America, see Roland Marchand, *Advertising the American Dream: Making Way for Modernity, 1920–1940* (Berkeley: University of California Press, 1985); T. J. Jackson Lears, *Fables of Abundance: A Cultural History of Advertising in America* (New York: Basic Books, 1994).

41. For a similar perspective, see Thomas F. Gieryn, "Science and Coca-Cola," *Science and Technology Studies* 5 (1987): 12–21.

42. John Burnham, *How Superstition Won and Science Lost* (New Brunswick, NJ: Rutgers University Press, 1987), 249, 244–45.

43. David Paul Nord has argued that while patterns in the mass media have often been assumed to mirror the popular mind, they more accurately reflect attempts to control markets and the interests of owners and publishers. See Nord, "An Economic Perspective on Formula in Popular Culture," *Journal of American Culture* 3 (1980): 17–31.

44. On the use of claims about science among professionalizing physicians, see Paul Starr, *The Social Transformation of American Medicine* (New York: Basic Books, 1982). For a general discussion of the uses of "scientific medicine" during the nineteenth century, see John Harley Warner, "The History of Science and the Sciences of Medicine," *Osiris* 10 (1995): 188–90. On the general relations between "science" and medicine during the late 1800s, see idem, "Ideals of Science and Their Discontents in Late Nineteenth-Century American Medicine," *Isis* 82 (1991): 454–78. On the uses of "science" among late-nineteenth- and early-twentieth-century engineers, see Ronald Kline, "Constructing 'Technology' as 'Applied Science': Public Rhetoric of Scientists and Engineers in the United States, 1880–1945," *Isis* 86 (1995): 194–221. On science in advertising during the 1920s, see Lears, *Fables of Abundance*, 124 and 174. For an early call for scientific advertising, see Harry Dexter Kitson, *Scientific Advertising* (New York: Codex Book Co., 1926).

45. On the early history of popular culture, see Peter Burke, *Popular Culture in Early Modern Europe* (New York: New York University Press, 1978).

46. John Fiske, *Understanding Popular Culture* (New York: Routledge, 1989), 25 and 43; idem, "Television: Polysemy and Popularity," *Critical Studies in Mass Communication* 3 (1986): 392. The capacity of popular culture to provide tools for keeping forces perceived as hegemonic at bay can also be related to David Hall's claim that the languages that make up popular culture are often playful. See Hall, "Introduction," 16.

47. To locate relevant magazine articles, I used a number of finding aids, including *Poole's Periodical Index* for the nineteenth century and *Readers' Guide to Periodical Literature* for the twentieth. I also relied on several online databases, including especially *Making of America*, a collaboration between Cornell University and the University of Michigan, and ProQuest Research Libraries' *Periodicals Contents*

Index. These databases are all accessible via the Internet at the following addresses: http://www.hti.umich.edu/m/moagrp (*Making of America*); http://poolesplus. odyssi.com/cgi-bin/phtml?ppsearch.htm (*Poole's Index to Periodical Literature*); http://vnweb.hwwilsonweb.com/hww/jumpstart.jhtml?prod=RDGFT,RGR (*Readers' Guide to Periodical Literature*); and http://pci.chadwyck.com (*Periodicals Contents Index*). *Making of America* includes the following magazines: *Appleton's Journal*, 1869–81 (2 series); *Catholic World*, 1865–1901; *DeBow's Review*, 1846–69 (3 series); *Garden and Forest*, 1888–97 (from Library of Congress); *Ladies' Repository*, 1841–76 (3 series); *Old Guard*, 1864; *Overland Monthly*, 1868–1900 (2 series); *Princeton Review*, 1831–82 (3 series); *Southern Literary Messenger*, 1835–64; *Southern Quarterly Review*, 1842–57 (3 series); *Vanity Fair*, 1860–62. It offers the advantage of in-text word searches among a variety of prominent nineteenth-century magazines from the 1830s to around 1900, including the *Princeton Review*, *Ladies' Repository*, *Catholic World*, and *Appleton's*. The available runs of these periodicals are neither coincident nor complete; some begin well after 1830 while others end before 1900. *Readers' Guide* and *Poole's Index* both cover hundreds of periodicals, but allow only title, subject, and keyword searches. They also include non-American publications, though their contents still provide important information about the ways in which English speakers were talking. None of these sources straddled the nineteenth and twentieth century to any significant extent, however. For a broad view, I relied on the *Periodicals Contents Index* (recently renamed *Periodicals Index Online*). Though again some of its listings are from places other than the United States, it offers thousands of periodicals of many kinds in which to search and a source of comparison with results from the other databases. In each case, I chose articles that addressed the four scientific controversies around which I have focused my study. I also examined secondary historical discussion of these four debates for citations of important articles and books. Finally, I followed references or citations in contemporary literature to sources that did not appear in either database or in the works of historians.

48. LaFollette, *Making Science*, 38.
49. Quoted in James Playsted Wood, *Magazines in the United States* (New York: Ronald Press, 1949), 42.
50. Ibid., 9.
51. Hillier Krieghbaum, *Science and the Mass Media* (New York: New York University Press, 1967), 61.
52. For a list of fiction that included references to phrenology, see John van Wyhe, "The History of Phrenology on the Web: Reading Phrenology," http://pages.britishlibrary.net/phrenology/literature.html (6 June 2005).

Chapter 1   Phrenology: A Science for Everyone

1. Nelson Sizer, *Forty Years of Phrenology* (New York: Fowler and Wells, 1882), 15–17, 158.
2. W.B.P., "A Letter from a Correspondent," *American Phrenological Journal*, 1 April 1839, 229.
3. Charles Dickens, *American Notes*, ed. Patricia Ingham (London: Penguin Books, 2000), 131.
4. John Davies, *Phrenology: Fad and Science* (New Haven, CT: Yale University Press, 1955), 55–56; Margaret Stern, *Heads & Headlines* (Norman: University of Oklahoma Press, 1971), 32.
5. "Phrenology," *Southern Literary Messenger*, March 1836, 286.

6. See, for instance, George Combe, "Moral and Intellectual Science: Natural Education," in *Moral and Intellectual Science* (New York: Fowler and Wells, 1848), 5–43; La Roy Sunderland, *Pathetism* (New York: P. P. Good, 1843), 13–14; "Neurology," *Western Quarterly Review*, January 1849, 66; "Literary Notices," *Ladies' Repository*, January 1849, 30. "Phrenological Observations on the Cerebral Development of William Burke and on the Development of William Hare," in *Illustrations of Phrenology*, ed. George H. Calvert (Baltimore, MD: W. and J. Neal, 1832), 71.

7. "Phrenology and Its Scientific Claim," *American Phrenological Journal*, January 1845, 7–8.

8. "Review of a Lecture on Phrenology by Frank H. Hamilton," *American Phrenological Journal*, October 1842, 299.

9. "Examination of Phrenology," *Southern Literary Messenger*, November 1839, 742; David Reese, *Humbugs of New-York* (New York: J. S. Taylor, 1838), 63.

10. This is based on a survey of 178 books and periodical articles that discussed phrenology between the 1830s and 1850s. Astronomy was mentioned eleven times, chemistry seven times, and geology six times. Except for anatomy, which was deeply implicated in debates over phrenology, all other branches of science were mentioned only two times or fewer.

11. "Agricultural Chemistry," review of Justus Liebig, *Chemistry in Its Application to Agriculture and Physiology*, *Princeton Review*, October 1844, 508. See also Margaret W. Rossiter, *The Emergence of Agricultural Science: Justus Liebig and the Americans, 1840–1880* (New Haven, CT: Yale University Press, 1975).

12. Sydney Ross, "*Scientist*: The Story of a Word," *Annals of Science* 18 (1962): 66–71.

13. Review of the *The Power of the Soul Over the Body*, *Southern Literary Messenger*, September 1847, 576; J.L.M., "The Women of France," *Southern Literary Messenger*, May 1839, 301.

14. Davies, *Phrenology*, 3–11.

15. On the *Vestiges*, see Secord, *Victorian Sensation*. On the radical community in Edinburgh and their relationship with materialistic science, see Adrian Desmond, *The Politics of Evolution* (Chicago: University of Chicago Press, 1989).

16. On Combe's bid for the Chair of Logic at the University of Edinburgh, see Gieryn, *Cultural Boundaries*, 116ff. On phrenology in Scotland and England generally, see Roger Cooter, *The Cultural Meaning of Popular Science* (Cambridge: Cambridge University Press, 1984); David deGiustino, *Conquest of Mind* (London: Croom Helm, 1975); G. N. Cantor, "The Edinburgh Phrenology Debate: 1803–1828," *Annals of Science* 32 (1975): 195–218; Steven Shapin, "Phrenological Knowledge and the Social Structure of Early Nineteenth-Century Edinburgh," *Annals of Science* 32 (1975): 219–43.

17. Benjamin Silliman, *American Journal of Science* 23 (1833): 356, quoted in Davies, *Phrenology*, 16.

18. "Obituary Notice of Dr. Gaspar Spurzheim," *American Journal of Science* 23 (1833): 368.

19. Charles Follen, *Funeral Oration* (Boston: Marsh, Capon and Lyon, 1832), 28, quoted in Davies, *Phrenology*, 25. For a partial list of American phrenological societies, see ibid., 27.

20. "Report on Infant Schools," *Annals of Phrenology* 1 (1834): 334–43. On phrenology in education, see Stephen Tomlinson, *Head Masters: Phrenology, Secular Education, and Nineteenth-Century Social Thought* (Tuscaloosa: University of Alabama Press, 2005).

21. Winter, "Orthodoxies and Heterodoxies"; idem, "Mesmerism and Popular Culture"; idem, *Mesmerized*, esp. 306–43.

22. On William Whewell, see Richard Yeo, *Defining Science: William Whewell, Natural Knowledge, and Public Debate in Early Victorian Britain* (Cambridge: Cambridge University Press, 1993). On the Gentlemen of Science and the BAAS, see Jack Morrell and Arnold Thackray, *Gentlemen of Science* (Oxford: Oxford University Press, 1981).

23. Alexander Dallas Bache, quoted in Hugh Slotten, *Patronage, Practice, and the Culture of American Science* (Cambridge: Cambridge University Press, 1994), 32. On Bache, see ibid. On Henry, see Albert E. Moyer, *Joseph Henry: The Rise of an American Scientist* (Washington, DC: Smithsonian Institution Press, 1997). On their desire to reform American science, see Slotten, *Patronage, Practice and Culture*, 25–41. On the Lazaronni and the formation of the AAAS, see Sally Gregory Kohlstedt, *The Formation of the American Scientific Community* (Urbana: University of Illinois Press, 1976). See also Robert Bruce, *The Launching of American Science* (Ithaca, NY: Cornell University Press, 1988); Nathan Reingold, *Science, American Style* (New Brunswick, NJ: Rutgers University Press, 1991), esp. 110–70.

24. Daniel Goldstein, "'Yours for Science': The Smithsonian Institution's Correspondents and the Shape of Scientific Community in Nineteenth-Century America," *Isis* 85 (1994): 577.

25. On informal botanists' motives, see Elizabeth Keeney, *The Botanizers* (Chapel Hill: University of North Carolina Press, 1992), 7 and 24. Also see Anne Secord, "Science in the Pub: Artisan Botanists in Early Nineteenth-Century Lancashire," *History of Science* 32 (1994): 268–315.

26. For biographical details for and a survey of the research challenges experienced by Joseph Henry, see, for example, Moyer, *Joseph Henry*.

27. Keeney, *Botanizers*, 29. For a general overview of the flow of scientific information during the early nineteenth century, see ibid., 25–36.

28. Sally Gregory Kohlstedt, "Parlors, Primers, and Public Schooling: Education for Science in Nineteenth-Century America," in *The Scientific Enterprise in America*, ed. Ronald L. Numbers and Charles E. Rosenberg (Chicago: University of Chicago Press, 1996), 61–82; Margaret W. Rossiter, "Benjamin Silliman and the Lowell Institute: The Popularization of Science in Nineteenth-Century America," *New England Quarterly* 44 (1971): 602–26; Stanley Guralnick, "Sources of Misconception on the Role of Science in the Nineteenth-Century American College," in *The Scientific Enterprise in America*, 83–97; idem, *Science and the Antebellum American College* (Philadelphia: American Philosophical Society, 1975); Donald Zochert, "Science and the Common Man in Ante-Bellum America," in *Science in America Since 1820*, ed. Nathan Reingold (New York: Science History Publications, 1976), 7–32. During the 1830s and 1840s, over 30 percent of articles indexed by *Making of America* (both the Cornell and University of Michigan versions) mentioned science. These searches were conducted in January 2003.

29. Frances Trollope, *Domestic Manners of the Americans* (New York; Vintage Books, 1960), 68; Davies, *Phrenology*, 57.

30. Mark Twain, *The Autobiography of Mark Twain*, ed. Charles Neider (New York: Harper, 1959), 64, quoted in Stern, *Heads & Headlines*, 20.

31. On the *American Journal of Phrenology*, see Davies, *Phrenology*, 62–63; Stern, *Heads & Headlines*, 64–67. On the Fowlers themselves, see Davies, *Phrenology*, 46–64; Stern, *Heads & Headlines*, 3–85.

32. Orson Fowler and Lorenzo Fowler, *Phrenology Proved, Illustrated and Applied*, 2d ed. (New York: Fowler and Wells, 1837), iii.

33. "Existing Evils and Their Remedy," *American Phrenological Journal*, February 1842, 41.

34. On Lydia Fowler's medical education and practice, see Stern, *Heads & Headlines*, 157–63. Her professorship was at the Central Medical College in Syracuse, NY. On the Fowler family's involvements with women's rights, see ibid., 166–71.

35. "Remarks on Phrenological Specimens, Cabinets, Etc.," *American Phrenological Journal*, February 1840, 213.

36. Davies, *Phrenology*, 38–39 and 47; Stern, *Heads & Headlines*, 22–23, 62–63.

37. Starr, *Transformation of American Medicine*, 30–59.

38. Nathan Hatch, *The Democratization of Christianity* (New Haven, CT: Yale University Press, 1989).

39. Starr, *Transformation of American Medicine*, 30–59.

40. Richard D. Brown, *Knowledge Is Power* (Oxford: Oxford University Press, 1989), 221.

41. Ibid., 13. On the general spread of information during the second quarter of the nineteenth century, see Secord, *Victorian Sensation*, esp. 41–154; Gerald Baldasty, *The Commercialization of News in the Nineteenth Century* (Madison: University of Wisconsin Press, 1992), 11–35; Isabelle Lehuu, *Carnival on the Page: Popular Print Media in Antebellum America* (Chapel Hill: University of North Carolina Press, 2000); John C. Nerone, *The Culture of the Press in the Early Republic: Cincinnati, 1793–1848* (New York: Garland, 1989); Ronald J. Zboray, *A Fictive People: Antebellum Economic Development and the American Reading Public* (Oxford: Oxford University Press, 1993); Brown, *Knowledge Is Power*, 218–21.

42. On attitudes to popularization, see Kohlstedt, *Formation*, 18–22; George H. Daniels, *American Science in the Age of Jackson* (New York: Columbia University Press, 1968), 40–41.

43. Winter, "Orthodoxies and Heterodoxies"; idem, "Mesmerism and Popular Culture"; idem, *Mesmerized*.

44. "Examination of Heads," *Annals of Phrenology* 2 (1835–36): 131; George Combe, quoted in Davies, *Phrenology*, 41; "Obituary Notice," 366; "Review of Hamilton," *American Phrenological Journal*, October 1842, 294.

45. Enoch Pond, "Phrenology," *Bibliotheca Sacra*, January 1854, 19.

46. On the nebular hypothesis, see Ronald L. Numbers, *Creation by Natural Law* (Seattle: University of Washington Press, 1977). On the tensions between Genesis and geology, see Charles Gillespie, *Genesis and Geology* (Cambridge, MA: Harvard University Press, 1951); James R. Moore, "Geologists and Interpreters of Genesis in the Nineteenth Century," in *God and Nature: Historical Essays on the Encounter between Christianity and Science*, ed. David C. Lindberg and Ronald L. Numbers (Berkeley: University of California Press, 1986), 322–27.

47. There is little work on the influence of Comte in the English-speaking world. For a discussion of Comte's interactions with prominent British intellectuals, see Diana Postlethwaite, *Making It Whole: A Victorian Circle and the Shape of Their World* (Columbus: Ohio State University Press, 1984).

48. For an overview of controversy over phrenology in the United States, see Davies, *Phrenology*, 65–78.

49. On the term "scientist," see Ross, "*Scientist*," 70–75; Richard Yeo, "Scientific Method and the Rhetoric of Science in Britain, 1830–1914," in *The Politics and Rhetoric of the Scientific Method*, ed. John Schuster and Richard Yeo (Dordrecht: Reidel, 1986), 273.

50. For example, in 178 sources, Herschel, who wrote a number of treatises on the workings of science, including his *Discourse on the Study of Natural Philosophy*, was cited twice and only once as a methodological guide.

51. "Phrenology, in Reply to the *Christian Examiner*," *New England Magazine*, March 1835, 182–83.

52. *Manual of Phrenology* (Philadelphia: Carey, Lea & Blanchard, 1835), 48–49; R. H. Collyer, *Manual of Phrenology* (Dayton, OH: B. F. Ellis, 1838), 7.

53. "Introductory Statement," *American Phrenological Journal*, 1 October 1838, 3–4.

54. "Miscellaneous," *American Phrenological Journal*, February 1854, 60.

55. Thomas Sewell, *Examination of Phrenology* (Washington, DC: Morgan, 1837), 62; "Sister Agnes," *Southern Literary Messenger*, April 1839, 273.

56. On attitudes to popularization, see Kohlstedt, *Formation*, 18–22; Daniels, *Age of Jackson*, 40–41.

57. "Introductory Statement," 3–4; "Phrenology in France," *American Phrenological Journal*, 1 December 1838, 74–75.

58. Fowlers, *Proved, Illustrated, Applied*, 421.

59. Andrew Boardman, *A Defence of Phrenology* (New York: Fowler and Wells, 1851), 20–60 and iv.

60. Orson Fowler, *Fowler on Memory* (New York, Fowler and Wells, 1842), 16.

61. "The Bearing and Influence of Phrenology," *American Phrenological Journal*, February 1844, 39.

62. Davies, *Phrenology*, 52.

63. Boardman, *Defence*, 17–18; "Phrenology," *American Journal of Science* 39 (1840): 70.

64. "President Shannon's Address," *American Phrenological Journal*, July 1839, 369.

65. "Combe on Phrenology, I," *Southern Literary Messenger*, June 1839, 394; Andrew Combe, "Phrenology: Its Nature and Uses," 64. George Combe describes his own conversion to phrenology in "Combe on Phrenology, I," 394.

66. Charles Caldwell, *Phrenology Vindicated* (New York, Samuel Colman, 1838), 26; "Miscellaneous," *American Phrenological Journal*, February 1842, 60.

67. "Notices," *Ladies' Repository*, January 1847, 30; Nathan Reingold, "Joseph Henry on the Scientific Life," in *Science, American Style*, 156–68; Slotten, *Patronage, Practice, and Culture*, 28–32.

68. For a history of the fact, see Mary Poovey, *A History of the Modern Fact* (Chicago: University of Chicago Press, 1998).

69. George S. Weaver, *Lectures on Mental Science* (New York: Fowler and Wells, 1852), 35; Nathaniel Bradstreet Shurtleff, *An Epitome of Phrenology* (Boston: March, Capen & Lyon, 1835), 8.

70. "Combe on Phrenology, II," *Southern Literary Messenger*, July 1839, 464.

71. George Combe, "On the Nature of the Evidence by Which the Functions of Different Parts of the Brain May Be Estimated," *American Phrenological Journal*, July 1839, 370, reprinted from *Edinburgh Phrenological Journal*.

71. W. Byrd Powell, *The Natural History of Human Temperaments* (Cincinnati: W. H. Derby, 1856), 203; Fowlers, *Proved, Illustrated, Applied*, 44.

73. Paul M. Roget, *Outlines of Physiology* (Philadelphia: Lea and Blanchard, 1839), 35 and 487; Dr. Thomson, "Phrenology," *Ladies' Repository*, December 1841, 366; "Examination of Phrenology," *Southern Literary Messenger*, November 1839, 742–47.

74. Orson Fowler, *Fowler's Practical Phrenology* (New York: Fowler and Wells, 1848), 8; Boardman, *Defence*, 10–11; "Miscellaneous," *American Phrenological Journal*, November 1839, 94–95, reprinted from the *Literary Examiner and Western Monthly Review*.

75. Francis Bacon, *New Organon*, quoted in Ernan McMullin, "Conceptions of Science in the Scientific Revolution," in *Reappraisals of the Scientific Revolution*, ed. David C. Lindberg and Robert S. Westman (Cambridge: Cambridge University Press, 1990), 45; Barry Gower, *Scientific Method* (London: Routledge, 1997), 40–62; Larry Laudan, *Science and Hypothesis* (Dordrecht: D. Reidel, 1981), 9–10; "Education in Europe," *Southern Literary Messenger*, January 1845, 3.

76. H. C. Foote, "Lecture of Phrenology," *American Phrenological Journal*, July 1858, 4.

77. Orson Fowler, *Phrenology Defended* (New York: O. S. and L. N. Fowler, 1842), 14; idem, *Memory and Intellectual Improvement* (New York: Fowler and Wells, 1847), 158–59; "Review of a Lecture by Frank H. Hamilton," *American Phrenological Journal*, 1 March 1842, 92.

78. Fowlers, *Proved, Illustrated, Applied*, v; Lydia Fowler, *Familiar Lessons on Phrenology*, vol. 2 (New York: Fowler and Wells, 1848), 176.

79. "Phrenological Examinations," *Cincinnati Mirror*, reprinted in *Southern Literary Messenger*, January 1835, 204; "Utility of Phrenology, cont.," *American Phrenological Journal*, 1 March 1839, 174.

80. Orson Fowler, *Matrimony* (New York: Fowler and Wells, 1851), x.

81. Nelson Sizer and Paul L. Buell, *Guide to Phrenology* (Woodstock, VT: Haskell and Palmer, 1842), 33–34.

82. Morrell and Thackray, *Gentlemen of Science*, 271. Richard Yeo has particularly stressed the publicly accessible nature of a Baconian method. See Yeo, "Scientific Method." For other works on Baconianism in American method talk, see Daniels, *Age of Jackson*, 63–101.

83. As in other cases, this is based on a survey of 178 phrenological sources from the 1830s, 1840s, and 1850s.

84. "Combe on Phrenology, III," *Southern Literary Messenger*, August 1839, 570; Calvert, *Illustrations*, vi.

85. T.C.C., "Morell's Argument against Phrenology," *American Whig Review*, August 1850, 195; Orson Fowler, "Practical Phrenology Defended," *American Phrenological Journal*, 1 September 1841, 576; J. Stanley Grimes, *Compendium of Phreno-Philosophy* (Boston: J. Monroe, 1850), 11; W.B.P., "A Letter," 229.

86. On Baconianism in religion, see Theodore Dwight Bozeman, *Protestants in an Age of Science: The Baconian Ideal and Antebellum American Religious Thought* (Chapel Hill: University of North Carolina Press, 1977); Herbert Hovenkamp, *Science and Religion in America, 1800–1860* (Philadelphia: University of Pennsylvania Press, 1978); George M. Marsden, "Everyone One's Own Interpreter? The Bible, Science and Authority in Mid-Nineteenth-Century America," in *The Bible in America*, ed. Nathan O. Hatch and Mark A. Noll (Oxford: Oxford University Press, 1982), 79–100. On Baconianism in art, see Jonathan Smith, *Fact and Feeling: Baconian Science and the Nineteenth-Century Literary Imagination* (Madison: University of Wisconsin Press, 1994); idem, "Art and Science," in *Scientific Methods: Conceptual and Historical Problems*, ed. Peter Achinstein and Laura J. Snyder (Malabar, FL: Krieger, 1994), 119–36. On Shakespeare as an example of comparison, see Orson Fowler, *Memory and Intellectual Improvement*, 157.

87. "The Sabbath Proved," *American Phrenological Journal*, June 1846, 176.

88. J.D.W., "Phrenology: A Socratic Dialogue," *American Whig Review*, January 1846, 43.

89. For some examples, see Moore, "Geologists and Interpreters"; Morgan B. Sherwood, "Genesis, Evolution, and Geology in America before Darwin: The Dana-Lewis Controversy, 1856–1857," in *Toward a History of Geology*, ed. Cecil J. Schneer (Cambridge, MA: MIT Press, 1969), 304–16; Stanley M. Guralnick, "Geology and Religion before Darwin: The Case of Edward Hitchcock, Theologian and Geologist (1793–1864)," *Isis* 63 (1972): 529–43; Philip J. Lawrence, "Edward Hitchcock: The Christian Geologist," *Proceedings of the American Philosophical Society* 116 (1972): 21–34.

90. James R. Moore has used this term to characterize the role of Baconian ideas in ensuring harmony between natural and divine realms. See Moore, "Geologists and Interpreters."

91. "Miscellaneous," *American Phrenological Journal*, November 1839, 94–95.
92. Mariano Cubi i Soler, *Phrenology* (Boston: Marsh, Capon, Lyon, and Webb, 1840), 102.
93. "President Shannon's Address," *American Phrenological Journal*, 1 July 1839, 370.
94 See, for instance, George Combe, "Moral and Intellectual Science: Natural Education," in *Moral and Intellectual Science*, 5–43; idem, "The Relation between Religion and Science," in *Moral and Intellectual Science*, 113–65.
95. William Scott, *Harmony of Phrenology with Scripture* (Edinburgh: Fraser & Co., 1836), 265.
96. H. C. Foote, "Is Phrenology Demoralizing?" *American Phrenological Journal*, December 1853, 127.
97. J. F. Graff, "Is Phrenology Profitable?" *American Phrenological Journal*, August 1852, 27.
98. Caleb Ticknor, "Popular Objections to Phrenology," *Knickerbocker*, April 1839, 308.
99. *American Phrenological Journal*, October 1844, 117.
100. Keeney, *Botanizers*, 101–7.
101. "Utility of Phrenology," *American Phrenological Journal*, February 1839, 147. Victor Hilts makes a similar point about analogy and faith in the universality of natural law. See "Toward the Social Organism: Herbert Spencer and William B. Carpenter on the Analogical Method," in *The Natural Sciences and the Social Sciences*, ed. I. B. Cohen (Boston: Kluwer Academic, 1994), 278. On analogy and natural theology, see Daniels, *Age of Jackson*, 178–83.
102. M.B., "Fallacy of Some Common Objections against Phrenology," *American Phrenological Journal*, 1 March 1839, 176; Orson Fowler and Lorenzo Fowler, *New Illustrated Self-Instructor in Phrenology* (New York: Fowler and Wells, 1859), 65–66; Orson Fowler, *Fowler on Memory* (New York: O. S. and L. N. Fowler, 1842), 80.
103. Robert MacNish, quoted in Roswell Wilson Haskins, *History and Progress of Phrenology* (Buffalo, NY: Steele & Peck, 1839), 207–8; the same quote appears in Boardman, *Defence*, 34–35; Rev. Dr. Drummond at a public dinner, quoted in Carmichael, *A Memoir on the Life and Philosophy of Spurzheim* (Dublin: W. F. Wakeman, 1833), 38–39; "Thoughts on Materialism, Insanity, Idiocy, Comparative Anatomy, Memory, Consciousness, &c.," *Annals of Phrenology* 2 (1835): 51.
104. "Phrenology and Revelation," *Methodist Quarterly Review*, April 1847, 168–69.
105. "Phrenology and Fact," *Methodist Quarterly Review*, October 1847, 585.
106. Reese, *Humbugs*, vi and viii.
107. Lorenzo Fowler, *Marriage* (New York: Fowler and Wells, 1847), 77.
108. Sizer, *Forty Years*, 93.
109. Fowlers, *Phrenology Proved, Illustrated, Applied*, 256–57; review of *The Positive Philosophy*, *Princeton Review*, June 1856, 69; *Christian Phrenology, a Book for the Million* (London: A. Drewett & Co., 1839), 16; "Biography of Dr. Gall," *American Phrenological Journal*, 1 October 1839, 1; Pond, "Phrenology," 643; "Phrenology," *Princeton Review*, April 1838, 283; Thomson, "Phrenology," 361; Lydia Fowler, *Familiar Lessons*, 22–24.
110. George Combe, *Constitution of Man* (Edinburgh: John Anderson, 1828), 43–44; Sizer and Buell, *Guide*, 28.
111. On Newton's apple, see, for example, Thomson, "Phrenology," 361; J. S. Allen, "Phrenology Examined," *Southern Literary Messenger*, May 1846, 267 and 277. On discussion of the story of the single bone, see Aldi Borondi Fasca Phorniostious, "The Aesthetics of Boots, No. III," *Vanity Fair*, 15 December 1860, 129; "Review of a Lecture on Phrenology by Frank H. Hamilton," 91.

112. On Franklin, see, for example, Josiah Graves, *A Phrenological Chart* (Norwich, CT: M. B. Young, 1839), 13; Calvert, *Illustrations*, 30.

113. "Phrenology," *American Journal of Science*, 87; on the long trial of true science, see Allen, "Phrenology Examined," 271; S.G.H., "The Heads of Our Great Men," *American Monthly Magazine*, April 1838, 355; Haskins, *History and Progress*, 29–60; Josiah M. Graves, *Lectures* (Norwich, CT: M. B. Young, 1838), 90–91; Reese, *Humbugs*, 18.

114. See, for instance, "Phrenological Examinations," 204; Joseph Buchanan, *Outlines of Lectures* (Cincinnati: Journal of Man, 1854), 53; Thomson, "Phrenology," 366.

115. "Phenology and Fact," 557; L. M. Lawson, "Popular Delusions," *Ladies' Repository*, February 1844, 49; "Phrenological Examinations," 204; Sizer, *Forty Years*, 237.

116. Susan Faye Cannon, *Science in Culture* (New York: Science History Publications, 1978), 2–3.

117. Ibid.; Yeo, *Defining Science*, 31.

118. Broman, "Habermasian Public Sphere," 142.

119. Daniels, "Professionalization," 31.

120. George Combe, introduction to *Moral and Intellectual Science*, 4.

121. John Fletcher, *The Mirror of Nature* (Boston: Cassady and March, 1839), 10–11; "Physical and Moral Science," *American Phrenological Journal*, October 1840, 15; "Dates Worth Remembering," *American Phrenological Journal*, June 1860, 88.

122. George Lakoff, *Women, Fire, and Dangerous Things: What Categories Reveal about the Mind* (Chicago: University of Chicago Press, 1987), 30–32.

123. From 1830 to 1859 the ratio between articles about science and articles with science in their titles was 0.04 (*Poole's*) or 0.036 (*PCI*). The ratio between articles with science in their texts and science in their titles over the same period was 0.01 (MOA). The first three of these databases show an increase in the fraction of articles about science and mentioning science that include science in their titles. From 1870 to 1899 the last two ratios were 0.25 (*PCI*) and 0.027 (MOA). *Poole's* shows a slight rise in the title/subject and title/text ratio. For the twentieth century, the subject to title ratio was much higher: 0.72 according to *Readers' Guide* and 0.37 according to *PCI*. These searches were conducted in January 2003.

124. General preference for discipline-specific labels over "man of science" was examined by searching the full texts of articles indexed by *Making of America*. "Scientist" is almost never used before mid-century, though by the latter portion of the 1800s, it had begun to eclipse others identifications, such as "man of science," "chemist," "astronomer," and "geologist." "Man of science" did not decline much in usage as the century progressed, but it was never more common than discipline-specific terms. These searches were conducted in January 2003.

125. Boardman, *Defence*, 16 and 20.

126. Stern, *Heads & Headlines*, 196. For a general discussion of the Fowlers and phrenology after the Civil War, see ibid., 181–259.

127. Ibid., 180.

128. J. Stanley Grimes, for example, who began writing on phrenology and phreno-mesmerism in the 1840s, had moved on to geology in 1893. See Grimes, *Essays on the Problems of Creation* (Philadelphia: Lippincott, 1893). Joseph Rodes Buchanan, who had begun writing about phreno-mesmerism during the 1840s, had moved on to psychic phenomena by the end of the century. See Buchanan, *Manual of Psychometry* (Boston: Holman Brothers, 1885).

Chapter 2   Evolution: Struggling over Science

1. George Axford, "Questions and Correspondence," Old and New, December 1872, 655–56, 658, 662.
2. "Evolution: What It Is Not, and What It Is," Popular Science Monthly, March 1888, 636.
3. Robert S. Ball, "The Relation of Darwinism to Other Branches of Science," Eclectic Magazine, December 1883, 816; Anna B. McMahan, "Recent Books on Evolution," Dial, May 1890, 7.
4. John Fiske, "The Progress from Brute to Man," North American Review, October 1973, 252.
5. J. W. Dawson, "The Present Aspect of Inquiries as to the Introduction of Genera and Species in Geological Time," Canadian Monthly, August 1872, 154.
6. Joseph LeConte, "Evolution in Relation to Materialism," Princeton Review, January-June 1881, 151.
7. "Darwin on Expression," Littell's Living Age, 5 July 1873, 4.
8. Axford, "Questions and Correspondence," 565.
9. Ibid., 657.
10. On Lamarck, see Richard W. Burkhardt, The Spirit of System: Lamarck and Evolutionary Biology (Cambridge, MA: Harvard University Press, 1977). On Enlightenment ideas about evolution, see Peter J. Bowler, Evolution: The History of an Idea (Berkeley: University of California Press, 1989), 50–89.
11. See Adrian Desmond, The Politics of Evolution (Chicago: University of Chicago Press, 1989).
12. On the Vestiges, see Bowler, Evolution, 141–50; Secord, Victorian Sensation. On Wallace's interests in phrenology, see Cooter, Cultural Meaning of Popular Science, 161 and 212. On his interest in the Vestiges, see Secord, Victorian Sensation, 332–33. On Wallace's religious and political ideas, also see Frank M. Turner, Between Science and Religion: The Reaction to Scientific Naturalism in Late Victorian England (New Haven: Yale University Press, 1974), 68–103.
13. Philip J. Lawrence, "Edward Hitchcock: The Christian Geologist," Proceedings of the American Philosophical Society 116 (1972): 32–34. Beyond occasional references, there is very little work on the reception of the Vestiges in America. Virtually the only work dedicated to this topic is Ryan Cameron MacPherson, "The Vestiges of Creation and America's Pre-Darwinian Evolution Debates: Interpreting Theology and Natural Sciences in Three Academic Communities" (Ph.D. diss., University of Notre Dame, 2003).
14. Joel S. Schwartz, "Darwin, Wallace, and Huxley, and Vestiges of the Natural History of Creation," Journal of the History of Biology 23 (Spring 1990): 149.
15. Asa Gray, review of Explanations, North American Review, April 1846, 478 and 505–6.
16. Scholars such as Gieryn and Frank Turner have linked the growing influence of a select group of professionalizing practitioners to visions of an independent and autonomous science. These visions stressed differences with other pursuits in order to heighten the value of scientific knowledge and work and to prevent meddling from perceived outsiders. See Gieryn, "Boundary-Work"; Gieryn, Bevins, and Zehr, "Professionalization of American Scientists"; Turner, "The Victorian Conflict." Links between professionalization and the removal of science from a common culture over the mid- to late-nineteenth century, a process that George Daniels calls "preemption," has also played a prominent role in Daniels, "Professionalization"; Keeney, Botanizers; Kohlstedt, American

Scientific Community; Starr, *Transformation of American Medicine*; Moyer, *Scientist's Voice*; Kevles, *Physicists*.

17. Keeney, *Botanists*, 131–32.
18. A. Hunter Dupree, *Science in the Federal Government* (Cambridge, MA: Harvard University Press, 1957); Thomas G. Manning, *Government in Science: The U.S. Geological Survey, 1864–1894* (Lexington: University of Kentucky Press, 1967); Robert E. Kohler, "The Ph.D. Machine: Building on the Collegiate Base," in *The Scientific Enterprise in America*, 98–122; Bruce, *Launching of Modern Science*, 314; John Lankford, *American Astronomy: Community Careers and Power, 1859–1940* (Chicago: University of Chicago Press, 1997), 76. For a detailed examination of the changing practices of American physics, see Daniel Kevles, *The Physicists* (Cambridge, MA: Harvard University Press, 1995), esp. 3–74.
19. On Darwin and his work, see Peter J. Bowler, *Charles Darwin: The Man and His Influence* (Cambridge: Cambridge University Press, 1996); Janet Browne, *Charles Darwin*, vol. 1 (London: Jonathan Cape, 1995); Adrian Desmond and James Moore, *Darwin* (London: Penguin Books, 1992).
20. For some references to Darwin, Huxley, and Tyndall (frequently in that order), see "Books and Authors," *Appleton's*, 17 June 1876, 796; Francis Bawon, "The Human and Brute Mind," *Princeton Review*, January-June 1880, 321; "Editor's Table," *Appleton's*, 18 April 1874, 504; James McCosh, "Contemporary Philosophy: Historical," *Princeton Review*, January-June 1878, 206; "The Darwinian Theory," *Atlantic Monthly*, October 1866, 425. On Huxley, see Paul White, *Thomas Huxley: Making the 'Man of Science'* (Cambridge: Cambridge University Press, 2002); Adrian Desmond, *Huxley: From Devil's Disciple to Evolution's High Priest* (Reading, MA: Addison-Wesley, 1997). On Tyndall, see William H. Brock, Norman D. McMillan, and R. Charles Mollan, *John Tyndall: Essays on a Natural Philosopher* (Dublin: Royal Dublin Society, 1981). On Darwin's defenders among the so-called X Club, see Ruth Barton, "'Huxley, Lubbock, and Half a Dozen Others': Professionals and Gentlemen in the Formation of the X Club, 1851–1864," *Isis* 89 (1998): 410–44. On the initial activities of Darwin's circle to rally around him immediately after the publication of the *Origin*, see Desmond and Moore, 467–81.
21. Ellegard, *Darwin and the General Reader*, 25; "Editorial Miscellany," *Debow's Review*, October 1860, 491.
22. "Literary Notes," *Appleton's*, 20 May 1871, 596.
23. On Joseph LeConte, see Lester D. Stephens, *Joseph LeConte: Gentle Prophet of Evolution* (Baton Rouge: Louisiana State University Press, 1982).
24. E. L. Youmans, "Herbert Spencer and His Philosophical System," *Appleton's*, 24 June 1871, 733.
25. For a discussion of such imagery in magazines during the very late 1800s, see Matthew Schneirov, *The Dream of a New Social Order* (New York: Columbia University Press, 1994).
26. "Use and Abuse of the Novel," *Catholic World*, November 1872, 254.
27. Asa P. Lyon, "How to Deal with Skeptics," *Ladies' Repository*, May 1872, 334.
28. "Darwinism," *Scribner's*, July 1875, 348.
29. John A. Garraty, *The New Commonwealth* (New York: Harper & Row, 1968), 86–87; "Standardization of Women's Clothing," National Institutes of Standards and Technology Virtual Museum, 2 September 2005, http://museum.nist.gov/exhibits/apparel/history.htm (27 January 2006).
30. Rodgers, *Contested Truths*.
31. James D. Norris, *Advertising and the Transformation of American Society, 1865–1920* (New York: Greenwood Press, 1990), xvi, 1–11, 34–40. On the general changes in

American society and culture, see Robert H. Wiebe, *The Search for Order* (New York: Hill and Wang, 1967); Richard Hofstadter, *The Age of Reform* (New York: Vintage Books, 1955); Garraty, *The New Commonwealth*; Carl Smith, *Urban Disorder and the Shape of Belief* (Chicago: University of Chicago Press, 1995); Carol Nackenoff, *The Fictional Republic: Horatio Alger and American Political Discourse* (New York: Oxford University Press, 1994); Frederic C. Jaher, *A Scapegoat in the New Wilderness: The Origin and Rise of Anti-Semitism in America* (Cambridge, MA: Harvard University Press, 1994); Robert E. Weir, *Beyond Labor's Veil: The Culture of the Knights of Labor* (University Park, PA: Pennsylvania State University, 1996); Norris, *Advertising*.

32. See Winter, "Orthodoxies and Heterodoxies"; idem, "Mesmerism and Popular Culture"; idem, *Mesmerized*. See also Laurence Levine, *Highbrow/Lowbrow: The Emergence of a Cultural Hierarchy in America* (Cambridge, MA: Harvard University Press, 1988); idem, "William Shakespeare and the American People: A Study in Cultural Transformation," in *Rethinking Popular Culture*, 157–97.

33. "American Authorship," *Princeton Review*, July 1873, 538.

34. This increase in articles mentioning "science" after a slight dip around 1850 appeared in a January 2003 search of *Making of America* (in both forms). *Poole's* did not show such a dip among articles about science; rather, it indicated a constant increase from the early part of the century.

35. Frank Luther Mott, *A History of American Magazines*, vol. 2 (Cambridge, MA: Harvard University Press, 1957), 78–93.

36. Mott, *History of American Magazines*, vol. 3, 417–18, on Youmans and *Appleton's*. On *Popular Science Monthly*, see pp. 495–99.

37. Scientific news began appearing in the "Editor's Scientific Record" in 1870.

38. Bernard Lightman makes this point for the late nineteenth century. See Lightman, "'The Voices of Nature': Popularizing Victorian Science," in *Victorian Science in Context*, ed. Bernard Lightman (Chicago: University of Chicago Press, 1997), 187–88.

39. "Editor's Table," *Appleton's*, 10 May 1873, 637; "Contemporary Literature," *Princeton Review*, October 1873, 758.

40. Ronald Numbers, "The Responses of American Naturalists to Evolution," *Darwinism Comes to America* (Cambridge, MA: Harvard University Press, 1998), 30–33. On Asa Gray, see A. Hunter Dupree, *Asa Gray, 1810–1888* (Cambridge, MA: Harvard University Press, 1959). On Louis Agassiz, see Lurie, *Louis Agassiz*.

41. H. P. Blavatsky, *Isis Unveiled* (New York: J. W. Bouton, 1877), xxxi.

42. Joseph LeConte, "Evolution in Relation to Materialism," *Princeton Review*, January-June 1881, 160; idem, "The Relation of Evolution to Materialism," *Popular Science Monthly*, May 1888, 80–81; idem, "What Is Evolution?" *Popular Science Monthly*, October 1887, 734.

43. "The Descent of Man," *Catholic World*, January 1878, 502.

44. E. L. Youmans, "Spencer's Evolution Philosophy," *North American Review*, October 1879, 399.

45. David Starr Jordan, "Evolution: What It Is and What It Is Not," *Arena*, August 1897, 153.

46. David Starr Jordan, *The Factors in Organic Evolution* (1895), 5. He makes a similar point in Jordan, "Evolution: What It Is and What It Is Not," 153.

47. Youmans, "Spencer's Evolution Philosophy," 399.

48. George S. Morris, "Philosophy and Its Specific Problems," *Princeton Review*, January-June 1882, 212, 218; Editor, "The Continuity of Evolution," *Monist*, October 1891, 70; Robert Flint, "Philosophy as Scientia Scientiarum," *Princeton Review*, July-December 1878, 697.

49. George S. Morris, "Philosophy and Its Specific Problems," *Princeton Review*, January-June 1882, 231.
50. Both *PCI* and *Poole's* showed an increase in talk about "pseudo-science" in article titles toward the end of the century. The term did not appear at all in titles indexed by *Poole's* before 1880. There was a similar increase among articles with "pseudo-science" in their texts as indexed by *Making of America*. These searches were conducted in January 2003.
51. Thomas Henry Huxley, "Scientific and Pseudo-Scientific Realism," *The Nineteenth Century*, February 1887, 191–204; reprinted in *Popular Science Monthly*, April 1887, 789–803; idem, "Science and Pseudo-Science," *The Nineteenth Century*, March 1887, 481–98; reprinted in *Popular Science Monthly*, June 1887, 207–24.
52. "Socialism and Communism in 'The Independent,'" *Catholic World*, March 1878, 808. On other examples of the use of pseudo-science, see "Contemporary Literature," *Princeton Review*, July 1873, 560; "Literary Notes," *Appleton's*, 23 November 1872, 584; A. F. Hewitt, "Scriptural Questions, Part III," *Catholic World*, February 1887, 654; "Darwin on Expression," *Littell's Living Age*, 5 July 1873, 3.
53. "Regent's Park," *Appleton's*, 8 August 1874, 178; "Literary Notes," *Appleton's*, 28 October 1871, 500.
54. J. S. Lippincott, "The Critics of Evolution," *American Naturalist*, May 1880, 319.
55. "Evolution: What It Is Not, and What It Is," 637.
56. Lawrence Irwell, "What Evolution Teaches Us," *Eclectic Magazine*, November 1894, 608–9.
57. Joseph LeConte, "Agassiz and Evolution," *Popular Science Monthly*, November 1887, 18.
58. This was true of searches of article titles in *PCI* and *Poole's* and searches of article texts in *Making of America* (both forms) conducted in January 2003.
59. Paul Shorey, "An Evolutionist's Alarm," *Dial*, 1 August 1893, 66.
60. "Current Literature," *Overland Monthly*, January 1874, 104.
61. Frank Hill, "Gaps in Agnostic Evolution," *Eclectic Magazine*, December 1895, 767; "Literary," *Appleton's*, 25 July 1875, 118.
62. "Editor's Table," *Appleton's*, 10 May 1873, 637.
63. "Editor's Table," *Appleton's*, 10 May 1873, 637; "More about Darwin," *Catholic World*, August 1873, 649–50.
64. Andrew Dickson White, *History of the Warfare of Science with Theology* (New York: D. Appleton, 1896), 70.
65. David C. Lindberg and Ronald L. Numbers, introduction to *God and Nature: Historical Essays on the Encounter Between Christianity and Science*, ed. David C. Lindberg and Ronald L. Numbers (Berkeley: University of California Press, 1986), 14. For a survey of important works on science and religion during the nineteenth century, see ibid., 6–8. On science and religion in the debates over evolution, also see John Hedley Brooke, *Science and Religion: Some Historical Perspectives* (Cambridge: Cambridge University Press, 1991), 275–320; A. Hunter Dupree, "Christianity and the Scientific Community in the Age of Darwin," in *God and Nature*, 351–68; Numbers, *The Creationists*, 3–36; idem, *Darwinism Comes to America*; James R. Moore, *The Post-Darwinian Controversies: A Study of the Protestant Struggle to Come to Terms with Darwin in Great Britain and America, 1870–1900* (Cambridge: Cambridge University Press, 1979); Jon H. Roberts, *Darwinism and the Divine in America: Protestant Intellectuals and Organic Evolution, 1859–1900* (Madison: University of Wisconsin Press, 1988); Turner, *Between Science and Religion*; idem, "The Victorian Conflict between Science and Religion"; Neal Gillespie, *Charles Darwin and the Problem*

of Creation (Chicago: University of Chicago Press, 1979); David N. Livingstone, "Re-placing Darwinism and Christianity," in *When Science and Christianity Meet*, ed. David C. Lindberg and Ronald L. Numbers (Chicago: University of Chicago Press, 2003), 183–202.

66. This was true of searches of article titles in *PCI* and *Poole's* and searches of article texts in *Making of America* (both forms) conducted in January 2003.

67. Charles Hodge, *What Is Darwinism?* (New York: Scribner, Armstrong & Co., 1874), 134–35. On Hodge and science, see Ronald L. Numbers, "Charles Hodge and the Beauties and Deformities of Science," in *Charles Hodge Revisited: A Critical Appraisal of His Life and Work*, ed. John W. Stewart and James H. Moorhead (Grand Rapids, MI: Eerdmans, 2002), 71–101.

68 The following conclusions are based on an examination of 315 statements about the relationship between science and religion in magazine articles discussing evolutionary ideas.

69. Wilbur F. Crafts, "Christianity a Science, Not a Dream," *New Englander*, February 1888, 115.

70. "Christianity and Modern Science," *Ladies' Repository*, May 1868, 360.

71. John W. Dawson, "The Present Rights and Duties of Science," *Princeton Review*, July-December 1878, 673.

72. "The Darwinian Theory," *Atlantic Monthly*, October 1866, 419; LeConte, "The Relation of Evolution to Materialism," 79.

73. Also Lindberg and Numbers, introduction to *God and Nature*, 2–3.

74. J. S. Lippincott, "The Critics of Evolution," *American Naturalist*, May 1880, 412.

75. Neal Gillespie calls this position the third episteme, the first being a positivistic model of science and the second being a science that admitted miraculous creation. See Gillespie, *Charles Darwin and the Problem of Creation*.

76. Joseph LeConte, *Evolution and Its Relation to Religious Thought* (New York: D. Appleton, 1889), 273; idem, "The Relation of Evolution to Materialism," 83.

77. W. D. Le Sueur, "Evolution and the Destiny of Man," *Popular Science Monthly*, February 1885, 468.

78. Washington Gladden, "The New Evolution," *McClure's*, August 1894, 235.

79. "Immortality and Evolution," *New Englander*, September 1884, 707.

80. John Fiske, *Excursions of an Evolutionist* (New York: Houghton Mifflin, 1902, orig. 1883), 276.

81. "A Final Philosophy," *Catholic World*, February 1878, 611.

82. "Current Literature," *Overland Monthly*, March 1873, 292.

83. St. George Mivart, "Evolution and Christianity," *Cosmopolitan*, July 1892, 146–51.

84. John Tyndall, "The Belfast Address," in *Fragments of Science* (New York: A. L. Burt Co., 1919), 491.

85. "The Darwinian Theory of the Origin of Species," *New Englander*, October 1876, 603.

86. On naturalism, see Turner, *Between Science and Religion*. On Huxley's use of religious language, see Dupree, "Christianity and the Scientific Community," 366.

87. David Strauss, quoted in "Current Literature," *Overland Monthly*, February 1874, 194. On the decline of religious belief among some groups during the late nineteenth century, see Susan Budd, *Varieties of Unbelief* (London: Heineman, 1977); Owen Chadwick, *The Secularization of the European Mind in the 19th Century* (New York: Cambridge University Press, 1975); Turner, *Without God, Without Creed*.

88. H. W. Conn, "The Three Great Epochs of World Evolution," *Methodist Review*, November 1896, 885.

89. Henry Drummond, *The Lowell Lectures on the Ascent of Man* (New York: J. Pott & Co., 1894), 342.

90. Henry A. Nelson, "God in Human Thought," *Princeton Review*, October 1875, 673.

91. Augustine Hewitt, "The Coming International Scientific Congress of Catholics," *Catholic World*, January 1888, 468; "Miscellany," *Appleton's*, 29 March 1873, 438; Balfour Steward, "Science and a Future State," *Princeton Review*, January-June 1878, 542; Dawson, "Rights and Duties of Science," 682.

92. John Bascom, *Evolution and Religion, or Faith as a Part of a Complete Cosmic System* (New York: G. P. Putnam, 1897), 4.

93. Tayler Lewis, quoted in Morgan B. Sherwood, "Genesis, Evolution, and Geology in America before Darwin: The Dana-Lewis Controversy, 1856–1857," in *Toward a History of Geology*, ed. Cecil J. Schneer (Cambridge, MA: MIT Press, 1969), 305.

94 "Angela, Chapter V," *Catholic World*, November 1869, 167. This was true of searches of article titles in PCI and *Poole's* and searches of article texts in *Making of America* (both forms) conducted in January 2003.

95. Thomas Henry Huxley, "The Progress of Science," in *Methods and Results* (New York: Appleton, 1899), 61–62; William James, "Great Men, Great Thoughts, and the Environment," *American Magazine*, October 1880, 457.

96. Fiske, *Century of Science*, 24.

97. "The Immutability of Species, Part II," *Catholic World*, December 1869, 336.

98. Hoffman, *Sphere of Science*, 7.

99. J. W. Dawson, "The Present Aspect of Inquiry as to the Introduction of Genera and Species in Geological Time," *Canadian Monthly*, August 1872, 154; Henry Calderwood, "The Problems Concerning Human Will," *Princeton Review*, July-December 1878, 334.

100. "Science and Atheism," *Ladies' Repository*, March 1868, 211; "Origin of Civilization," 503; W. J. Beecher, "Faith as an Ambiguous Middle Term," *Princeton Review*, July 1873, 464; "Contemporary Literature," *Princeton Review*, January 1874, 176; "On the Higher Education," *Catholic World*, March 1871, 729.

101. Alexander Winchell, "Voices from Nature, Part VI," *Ladies' Repository*, July 1863, 390.

102. John Tyndall, "On the Scientific Use of the Imagination," *Appleton's*, 29 October 1870, 527.

103. Stanley Jevons, *Principles of Science* (London: Macmillan, 1874), 576; Frank Sargent Hoffman, *The Sphere of Science* (New York: G. P. Putnam's Sons, 1898), 103–4.

104. Drummond, *Lowell Lectures*, 9.

105. George F. Wright, "Recent Works Bearing on the Relation of Science to Religion," *Bibliotheca Sacra*, January 1880, 70–72; William Barry, "Prof. Huxley's Admissions, Part I," *Catholic World*, December 1894, 333; Annie Bessant, *Evolution of Life and Form* (Benares: Theosophical Publishing Society, 1905), 135–36.

106. "The True Evolution," *Arena*, June 1897, 1100; Jordan, "Evolution: What It Is and What It Is Not," 159.

107. Rev. G. M. Searle, "Evolution and Darwinism," *Catholic World*, November 1892, 226; "Charles Robert Darwin," *Popular Science Monthly*, February 1873, 497; David Starr Jordan, "Darwin," *Dial*, May 1882, 3; "Charles Darwin and Evolution," *Living Age*, 16 September 1882, 643, 646; E. Lawrence, "Modern Forms of Theistic Naturalism," *Ladies' Repository*, May 1870, 334; "New Publications," *Catholic World*, May 1884, 284.

108. "Character of Herbert Spencer," *Appleton's*, 23 October 1869, 309.

109. A. F. Hewitt, "Scriptural Questions, Part III," *Catholic World*, February 1887, 656; Noah Porter, "The New Atheism," *Princeton Review*, January-June 1880, 370; "Christianity and Modern Science," *Ladies' Repository*, May 1868, 361, 364; "The Physical Basis of Life," *Catholic World*, July 1869, 472.
110. "The Skepticism of Science," *Princeton Review*, January 1863, 52–53.
111. "Blue-Eyes and Battlewick, Chapters V-XI," *Southern Literary Messenger*, February 1860, 105.
112. "Books of the Day," *Appleton's*, December 1878, 576; "Contemporary Literature," *Princeton Review*, April 1876, 370.
113. "Dramatic Notes," *Appleton's*, 3 February 1872, 135.
114. Henry C. Carey, "Science and Its Methods," *Pennsylvania Monthly*, April 1872, 157–71; May 1872, 214.
115. E. L. Youmans, *The Culture Demanded by Modern Life* (New York: D. Appleton & Co., 1867), 373–74; "Current Literature," *Overland Monthly*, February 1873, 200.
116. Horace Bushnell, *Nature and the Supernatural* (New York: Charles Scribner, 1858), 20; "The Future of Human Character," *Ladies' Repository*, January 1868, 43.
117. George Romanes, "A Reply to the 'Fallacies of Evolution,'" *Popular Science Monthly*, November 1879, 103.
118. McDermot, "The Church and the New Sociology," 296.
119. E. L. Youmans, "Spencer's Evolution Philosophy," *North American Review*, October 1879, 392.
120. "Skepticism of Science," 143.
121. H. C. Alexander, "Reason and Redemption," *Princeton Review*, July 1875, 426.
122. See chapter 2, n.124.
123. "Motion by Rail," *Appleton's*, 8 June 1872, 634.
124. "Some Religious Studies and Speculations, Part II," *Overland Monthly*, April 1891, 443.
125. Sara A. Underwood, "Darwin and His Work," *Open Court*, 3 March 1887, 41.
126. "Miscellany," *Appleton's*, 25 March 1871, 358.
127. T. H. Huxley, *Darwiniana* (New York: D. Appleton, 1893), 363.
128. George Romanes, "A Reply to the 'Fallacies of Evolution,'" *Popular Science Monthly*, November 1879, 103; St. George Mivart, *An Introduction to the Elements of Science* (Boston: Little, Brown & Co., 1894), 2.
129. William H. Haskell, "Cremation," *Ladies' Repository*, July 1874, 40.
130. Thomas Hill, "Erasmus Darwin," *Bibliotheca Sacra*, July 1878, 481.
131. Edward B. Howell, "The Inductive Method and Religious Truth," *New Englander*, May 1891, 474–75.
132. Turner, "The Victorian Conflict between Science and Religion," 369.
133. Asa Gray, "Charles Darwin," *American Journal of Science*, December 1882, 460; Marion Hamilton Carter, "Darwin's Idea of Mental Development," *American Journal of Psychology*, July 1898, 536.
134. Louis Robinson, "Evolution and the Amateur Naturalist," *Eclectic Magazine*, July 1897, 28.
135. Bernard Lightman, "Constructing Victorian Heavens: Agnes Clerk and the 'New Astronomy,'" in *Natural Eloquence: Women Reinscribe Science*, ed. Barbara T. Gates and Ann B. Shteir (Madison: University of Wisconsin Press, 1997), 71–73.
136. "'Darwin Answered,'" *Pennsylvania Monthly*, May 1875, 368; Joseph LeConte, "Rough Notes of a Yosemite Camping Trip, Part III," *Overland Monthly*, December 1885, 635.
137. J. W. Dawson, "Evolution and the Apparition of Life Forms," *Princeton Rreview*, January-June 1878, 666; George McDermot, "The Church and the New Sociology,"

Catholic World, December 1895, 294; "The Literary Extravagance of the Day," *Catholic World*, May 1878, 257.

138. John W. Mears, "Theistic Reactions in Modern Speculation," *Princeton Review*, April 1875, 329.

139. "Three Lectures on Evolution," *Catholic World*, February 1877, 616.

140. Frederic Gardiner, "Darwinism," *Bibliotheca Sacra*, April 1872, 240.

141. Thomas Hill, "The Struggle of Science," *Unitarian Review*, April 1875, 335.

142. This is based on a survey of statements about those who practiced science in literature on evolution from 1860 to 1900. Assertions of the narrowness of scientists accounted for 46 cases. The next category (38 cases) included comments and concerns about ordinary people's access to scientific practice. After this came discussion of the ordinariness of those who did science (36), specialization (26), Darwin as a model practitioner (25), the role of practitioners in the diffusion of science (17), the division of labor in science (14), the machinelike personality of many men of science (12), the distinction between genius and industry (12).

143. Albin Putzker, "Letters and Science," *Overland Monthly*, July 1890, 17.

144. "Darwin's Life," *Atlantic Monthly*, April 1888, 560–66, 565.

145. Julia Scott, "Phoebus or Cupid," *Overland Monthly*, August 1886, 129–33.

146. "Contemporary Literature," *Ladies' Repository*, June 1871, 469.

147. Simon Newcomb, "Evolution and Theology," *North American Review*, June 1879, 663.

148. Moyer, *A Scientist's Voice*, 140.

149. Noah Porter, "The New Atheism," *Princeton Review*, January-June 1880, 370; John A. Mooney, "The World in a Drop of Water," *Catholic World*, June 1889, 356.

150. "The Evolution of Life," *Catholic World*, May 1873, 155.

151. Alfred Lord Tennyson, *The Poetic and Dramatic Works of Alfred Lord Tennyson*, ed. W. J. Rolfe (Boston: Houghton Mifflin, 1898), 176.

152. "The *Princeton Review* and Leo XIII," *Catholic World*, June 1880, 392; Charles Elam, "Automatism and Evolution," *Eclectic Magazine*, December 1876, 646; "Wallace's Darwinism," *Nation*, 71 October 1889, 356.

153. H.I.D. Ryden, "The Proper Attitude of Catholics toward Modern Biblical Criticism," *Catholic World*, June 1893, 403.

154. Leon C. Field, "The Old and the New," *Ladies' Repository*, February 1873, 93.

155. A. F. Hewitt, "A New Road from Agnosticism to Christianity," *Catholic World*, October 1895, 13; Editor, "The Message of Monism to the World," *Monist*, July 1894, 545; Henry Wood, "Omnipresent Divinity," *Arena*, September 1895, 77; Henry Drummond, *Natural Law in the Spiritual World* (New York: Lovell, Coryell, & Co., 1884), 20–21; "The Military Novel," *Catholic World*, May 1880, 161; Crafts, "Christianity a Science," 114.

156. Jon H. Roberts and James Turner, *The Sacred and Secular University* (Princeton, NJ: Princeton University Press, 2000).

157. E. F. Carr, "A Theory, an Extravaganza," *Ladies' Repository*, August 1873, 125.

158. LeConte, "Rough Notes," 635.

159. Alpheus Packard, "A Half-Century of Evolution," *Science*, 26 August 1898, 246.

160. Andrew Dickson White, quoted in Kevles, *Physicists*, 35.

161. John Fiske, "Agassiz and Darwin," *Popular Science Monthly*, October 1873, 695.

162. "Editor's Table," *Appleton's*, 29 January 1876, 152.

163. G. Stanley Hall, "The Moral and Religious Training of Children," *Princeton Review*, January-June 1882, 37–38.

164. E. L. Youmans, "The Burden of Knowledge," *Appleton's*, 11 September 1869, 120.

165. On the history of the evolutionary synthesis, see, for instance, Ernst Mayr and William B. Provine, eds., *The Evolutionary Synthesis: Perspectives on the Unification of Biology* (Cambridge, MA: Harvard University Press, 1980).

166. On the Scopes trial, see, for instance, Edward J. Larson, *Summer for the Gods: The Scopes Trial and America's Continuing Debate over Science and Religion* (New York: Basic Books, 1997). On the antievolution crusade of the 1920s and after, see idem, *Trial and Error: The American Controversy over Creation and Evolution* (New York: Oxford University Press, 1985); Numbers, *The Creationists*; idem, *Darwinism Comes to America*, 58–135.

Chapter 3 Relativity: A Science Set Apart

1. "Rioting for Science," *Scientific American*, March 1930, 188.
2. Charles J. Liebman, "The Thirst for Knowledge," *New York Times*, 15 January 1930, 24; H. Horton Sheldon, *Space, Time, and Relativity* (New York: University Society, 1932), 3; "Rioting for Science," 188.
3. Charles Steinmetz, *Four Lectures on Relativity and Space* (New York: McGraw-Hill, 1923), v.
4 Julian S. Huxley, "A Journey in Relativity," *North American Review*, July 1923, 67.
5. E. E. Slossen, *Easy Lessons in Einstein* (New York: Harcourt, Brace, and Howe, 1920), 1. On public reaction to relativity and Einstein, see David Cassidy, *Einstein and Our World* (Atlantic Highlands, NJ: Humanities Press, 1995), esp. 60–90; Marshall Missner, "Why Einstein Became Famous in America," *Social Studies of Science* 15 (1985): 267–91; Ronald C. Tobey, *The American Ideology of National Science* (Pittsburgh: University of Pittsburgh Press, 1971), 104–15.
6. Albert Einstein, "Personal God Concept Causes Science-Religion Conflict," *Science News Letter*, 21 September 1940, 181.
7. William F. Wunsch, "Relativity," *New-Church Review*, July 1931, 347.
8. Tobey, *American Ideology*, 229.
9. Viscount Haldane, *The Reign of Relativity* (New Haven, CT: Yale University Press, 1921), vii.
10. Oliver Lodge, *Relativity* (New York: George H. Doran, 1926), 7; Frederick E. Brasch, "Bibliography of Relativity," *Science*, 30 September 1921, 304; Henry Hazlitt, "Einstein," *Nation*, 19 November 1930, 554; Charles Lane Poor, *Gravitation versus Relativity* (New York: G. P. Putnam, 1922), iv; Haldane, *Reign*, 7.
11. Robert H. Wiebe, *Search for Order, 1877–1920* (New York: Hill and Wang, 1967), 113–27; Richard Hofstadter, *The Age of Reform: From Bryan to FDR* (New York: Knopf, 1955), 148–64; Starr, *Transformation of American Medicine*, 93–144.
12. On the development of the academic research and graduate training, see Roger L. Geiger, *To Advance Knowledge: The Growth of the American Research Universities, 1900–1940* (New York: Oxford University Press, 1986); Kevles, *Physicists*, 60–74 and 220–21; Kohler, "Ph.D. Machine"; Larry Owen, "MIT and the Federal 'Angel': Academic R&D and Federal-Private Cooperation before World War II," in *The Scientific Enterprise in America*, 247–72. On professional organizations, see Marc Rothenberg, "Organization and Control: Professionals and Amateurs in American Astronomy, 1899–1918," *Social Studies of Science* 11 (1981): 305–25; Thomas L. Haskell, *The Emergence of Professional Social Science: The American Social Science Association and the Nineteenth-Century Crisis of Authority* (Urbana: University of Illinois Press, 1977).
13. On amateur science in the early twentieth century, see Rothenberg, "Organization and Control," 305–25; John Lankford, "Amateurs and Astrophysics: A

Neglected Aspect in the Development of a Scientific Specialty," *Social Studies of Science* 11 (1981): 275–303.

14. On Einstein and his work of relativity, see Albrecht Folsing, *Albert Einstein* (New York: Viking Press, 1997); Banesh Hoffman, *Albert Einstein: Creator and Rebel* (New York: Plume, 1972); Gerald Holton, *Thematic Origins of Scientific Thought: Kepler to Einstein* (Cambridge, MA: Harvard University Press, 1982).

15. In the U.S., mathematician Robert Carmichael gave the first series of lectures on relativity in 1912. See Robert Carmichael et. al., *A Debate on the Theory of Relativity* (Chicago: Open Court, 1927), 2.

16. J. S. Aimes, "Einstein's Law of Gravitation," *Science*, 12 March 1920, 253. On physicists' reaction to relativity, see Arthur I. Miller, *Albert Einstein's Special Theory of Relativity: Emergence and Early Interpretation* (Reading, MA: Addison-Wesley, 1981); Richard Staley, "On Histories of Relativity: The Propagation and Elaboration of Relativity Theory in Participant Histories in Germany, 1905–1911," *Isis* 89 (1998): 263–99; Kevles, *Physicists*, 84–90; Loren R. Graham, "The Reception of Einstein's Ideas: Two Examples from Contrasting Political Cultures," in *Albert Einstein: Historical and Cultural Perspectives*, ed. Gerald Holton and Yehuda Elkana (Princeton, NJ: Princeton University Press, 1982); Thomas F. Glick, ed., *The Comparative Reception of Relativity* (Boston: Reidel, 1987).

17. Aimes, "Einstein's Law," 253.

18. See J. Leonard Bates, *The United States, 1898–1928: Progressivism and a Society in Transition* (New York: McGraw-Hill, 1976); Paul Carter, *The Twenties in America* (New York: Thomas Y. Crowell Co., 1968); Geoffrey Perret, *America in the Twenties* (New York: Simon & Schuster, 1982).

19. Tobey, *American Ideology*, 20–61; Hugh Slotten, "Human Chemistry or Scientific Barbarism? American Responses to World War I Poison Gas, 1915–1930," *Journal of American History* 77 (1990): 476–98; Kevles, *The Physicists*, 236–51; Peter J. Kuznick, *Beyond the Laboratory: Scientists as Political Activists in 1930s America* (Chicago: University of Chicago Press, 1987).

20. Brief discussion of scientific management makes frequent appearance in general historical work on the 1920s. For instance, see Noble, *America by Design*, 164–78; Samuel Haber, *Efficiency and Uplift: Scientific Management in the Progressive Era, 1890–1920* (Chicago: University of Chicago Press, 1964). On scientific motherhood, see Rima D. Apple, *Mothers and Medicine: A Social History of Infant Feeding, 1890–1950* (Madison: University of Wisconsin Press, 1987).

21. "Einstein and the Press," *New Republic*, 13 February 1929, 336.

22. Quoted in Kevles, *Physicists*, 98.

23. On the changing nature of the public health movement, see Tomes, *Gospel of Germs*, 237–55. Tomes does not mean to imply that enthusiasm for public health waned, but that it shifted direction during the 1920s toward more stringent, laboratory-based methods and away from the more informal and homely reform of earlier decades. On eugenics, see Diane B. Paul, *Controlling Human Heredity* (Amherst, NY: Humanity Books, 1995), 77–84; Daniel Kevles, *In the Name of Eugenics* (New Brunswick, NJ: Rutgers University Press, 1963); Edward J. Larson, *Sex, Race, and Science: Eugenics in the Deep South* (Baltimore, MD: Johns Hopkins University Press, 1995); Mark H. Haller, *Eugenics: Hereditarian Attitudes in American Thought* (New Brunswick, NJ: Rutgers University Press, 1963); Donald K. Pickens, *Eugenics and the Progressives* (Nashville: Vanderbilt University Press, 1968).

24. This was true of articles about science indexed by *Readers' Guide* and articles with "science" in the title indexed by *Readers' Guide* and PCI. These searches were conducted in January 2003.

25. Kevles, *Physicists*, 96. Kevles argues that during the early years of the twentieth century, scientific knowledge had become too technical for the layperson and science itself had become detached from the culture of generally educated Americans.
26. The creation of Science Service is treated in a number of places. See Kevles, *Physicists*, 171–72; LaFollette, *Making Science*, 10; Hillier Krieghbaum, *American Newspaper Reporting of Science News* (Manhattan, KA: Kansas State College of Agriculture and Applied Science, 1941), 49–61.
27. LaFollete, *Making Science*, 48.
28. Michael Schudson, *Discovering the News* (New York: Basic Books, 1978), 72.
29. See Marchand, *American Dream*; Lears, *Fables of Abundance*.
30. See, for instance, David J. Rhees, "Corporate Advertising, Public Relations, and Popular Exhibits: The Case of Du Pont," *History and Technology* 10 (1993): 67–75; Burnham, *How Science Lost*, 212–14; LaFollette, *Making Science*, 54–60. On the general growth of public relations, see Lears, *Fables of Abundance*, 203–4.
31. Henry Crew, "The Exposition of Science," *Science Monthly*, September 1932, 233.
32. Robert Morgan, "Einstein Theory on Trial," *Illustrated World*, September 1923, 33.
33. See, for example, C. A. Chant, "Eclipse Displacement on the Plates Taken by the Canadian Party at the Australian Eclipse," *Science*, 28 September 1923, supplement 10–13; "Professor Michelson's New Einstein Test," *Literary Digest*, 4 July 1925, 26; "New California Experiment Supports Einstein Theory," *Scientific American*, June 1927, 413; "Two to one in favor of Einstein," *Outlook*, 12 August 1925, 512; H. W. Carr, "The Principle of Relativity and the Dayton Miller Experiments," *Nineteenth Century*, October 1926, 556–68.
34. See, for example, "'Bolsheviki of Science' Who Follow Einstein," *Literary Digest*, 9 April 1921, 72–74; "Getting Back at Einstein," *Literary Digest*, 4 June 1921, 29–30; John T. Blankhart, "Relativity and Interdependence," *Catholic World*, February 1921, 588–610; "Is the Einstein Theory a Crazy Vagary?" *Literary Digest*, 2 June 1923, 29–30; "Is Einstein's Arithmetic Off?" *Literary Digest*, 8 November 1924, 20–21; "Captain See versus Doctor Einstein," *Scientific American*, February 1925, 128; J. Malcolm Bird, "An Alternative to Einstein: How Dr. Poor Would Save Newton's Laws and the Classical Time and Space Concepts," *Scientific American*, 11 June 1921, 468 and 479; "Criticizing Professor Einstein," *Scientific American*, December 1924, 131; "Going Back on Einstein," *Literary Digest*, 5 February 1927, 26.
35. Missner, "Why Einstein Became Famous," 268.
36. LaFollette, *Making Science*, 39.
37. The conclusion about the *New York Times* is based on a search of articles mentioning "science" indexed by ProQuest Historical Newspapers, http://proquest. umi.com/. Articles about science in *PCI* amounted to 4 percent of the total number of articles indexed during the 1870s; this percentage decreased over the last decades of the 1800s and never reached above 3 percent during the twentieth century. Similarly, articles indexed in *Readers' Guide* with "science" in their titles were never as frequent during the early 1900s, relative to the total number of articles indexed, as they were during the 1890s. These searches were conducted in January 2003.
38. This conclusion is based on a survey of advertisements in the *Saturday Evening Post* in randomly chosen issues from 1924 to 1927.
39. Tobey, *American Ideology*, 3.
40. "Einstein's Latest," 179.

41. "After All, Einstein Is a Human Being," 38–40; "Einstein and Religion," *Commonweal*, 29 October 1930, 653.
42. Tobey, *American Ideology*, 20–61; Hugh Slotten, "Human Chemistry or Scientific Barbarism? American Responses to World War I Poison Gas, 1915–1930," *The Journal of American History* 77 (1990): 476–98; Kevles, *Physicists*, 236–51; Peter J. Kuznick, *Beyond the Laboratory: Scientists as Political Activists in 1930s America* (Chicago: University of Chicago Press, 1987).
43. Schudson, *Discovering the News*, 150; Marchand, *American Dream*, 357. See also Joan Shelley Rubin, *The Making of Middle-Brow Culture* (Chapel Hill: University of North Carolina Press, 1992).
44. The interview was reported in the *New York Evening Post*. A summary appeared in "Einstein Finds the World Narrow," *Literary Digest*, 16 April 1921, 32–34; "We May Not 'Get' Relativity, But We Like Einstein," *Literary Digest*, 27 December 1930, 29–30; Albert Einstein, "What I Believe," *Forum*, June 1930, 193–94. The statement on pacifism appeared in the *New York Times*. An extended quote also appeared in "Einstein on Peace," *Nation*, 5 February 1930, 144.
45. *Nation*, 14 December 1932, 58.
46. For a spoof on these difficulties, see Eccentricus, "Saving America," *World Tomorrow*, 21 December 1932, 584.
47. Portraits of Einstein appeared in *Women's Journal*, January 1931, 16; *Business Week*, 22 February 1936, 9; *Fortune*, February 1936, 78; *Scholastic*, 3 October 1936, 15. See also "Tagore Talks with Einstein," *Asia*, March 1931, 138–42. Quote is from "Einstein on Peace," 133. See also "Einstein's Peace Appeal," *Nation*, 2 September 1931, 239; "The 1932 Disarmament Conference," *Nation*, 23 September 1931, 300. Also see "Einstein the Pacifist," *Christian Century*, 24 December 1930, 1580–81; "Einstein, Physicist and Pacifist," *Christian Century*, 18 March 1931, 363–64; "Einstein Is Now an American," *Christian Century*, 16 October 1940, 128.
48. Hansen, "America's First Medical Breakthrough."
49. "Einstein's Cosmic Religion," *Literary Digest*, 29 November 1930, 19; "Why Einstein's Theory Is So Hard to Explain," *Current Opinion*, July 1920, 77; "Has Einstein Turned Physics into Metaphysics?" *Current Opinion*, June 1921, 802.
50. Editor's note to E. E. Slossen, "Can You Tell the Difference between Rest and Motion?" *Independent*, 13 December 1919, 174.
51. Poor, *Gravitation versus Relativity*, vii.
52. Albert Einstein and Leopold Infeld, *The Evolution of Physics* (New York: Simon & Schuster, 1938), 159.
53. "Professor Einstein, Famous Revisor of the Universe, 'at Home,'" *Literary Digest*, 20 March 1920, 73.
54. "Professor Einstein at the California Institute of Technology," *Science*, 10 April 1931, 376.
55. Sheldon, *Space, Time, and Relativity*, 2.
56. Hans Reichenbach, *From Copernicus to Einstein* (New York: Philosophical Library, 1942), 11.
57. E. E. Slossen, "That Elusive Fourth Dimension," *Independent*, 27 December 1919, 274–75 and 296–98. On the general usage of the idea of revolution in the history of science, see I. Bernard Cohen, *Revolution in Science* (Cambridge: Harvard University Press, 1985); David C. Lindberg, "Conceptions of the Scientific Revolution from Bacon to Butterfield: A Preliminary Sketch," in *Reappraisals of the Scientific Revolution*, ed. David C. Lindberg and Robert S. Westman (New York: Cambridge University Press, 1990), 1–21.
58. Levi Gruber, *The Einstein Theory* (Burlington, IA: Lutheran Literary Board, 1923), 17.

59. Charles Lane Poor, "What Einstein Really Did," *Scribner's*, November 1930, 527.

60. Axel Idestrom, *The Relativity Theories of Einstein—Untenable* (Uppsala: Almqvist & Wiksells, 1948), 16.

61. Paul R. Heyl, "Space, Time and Einstein," *Science Monthly*, September 1929, 234.

62. John A. Eldridge, "The Awkward Consistency of Science," *Science Monthly*, March 1929, 363.

63. Authre Stanley Eddington, *The Nature of the Physical World* (Cambridge: Cambridge University Press, 1929), 282.

64. Eddington, *Space, Time and Gravitation*, 56.

65. Morgan, "Einstein's Theory on Trial," 35–36.

66. On concern over vivisection, see Susan E. Lederer, "The Controversy over Animal Experimentation in America, 1880–1914," in *Vivisection in Historical Perspective*, ed. Nicholaas A. Rupke (London: Croom Helm, 1987), 235–58; Patricia Peck Gossel, "William Henry Welch and the Antivivisection Legislation in the District of Columbia, 1896–1900," *Journal of the History of Medicine and Allied Sciences* 40 (1985): 397–419. On concern over human experimentation, see Susan E. Lederer, "'The Right and Wrong of Making Experiments on Human Beings': Udo J. Wile and Syphilis," *Bulletin of the History of Medicine* 58 (1984): 380–97; idem, *Subjected to Science: Human Experimentation in America before the Second World War* (Baltimore, MD: Johns Hopkins University Press, 1995). On the debate over the use of poison gas in World War I, see Slotten, "Human Chemistry or Scientific Barbarism?" esp. 481–83.

67. Schudson, *Discovering the News*, 123.

68. Ibid., 135. See also Ellen Fitzpatrick, ed., *Muckraking: Three Landmark Articles* (Boston: Bedford Books of St. Martin's Press, 1994), 12–22 and 103–6; Marchand, *Advertising the American Dream*. For a more in-depth account of the changes in journalism during this period, see Schudson, *Discovering the News*, 121–59.

69. "The Row among the Physicists," *Nation*, 27 December 1919, 819.

70. "Einstein Pays a Call," *Outlook*, 3 December 1930, 529.

71. "Einstein and the Press," 227.

72. This was true of articles with "scientific method" in their titles as indexed by both *PCI* and *Readers' Guide*. These searches were conducted in January 2003.

73. W. C. Croxton, *Science in the Elementary School* (New York: McGraw-Hill, 1937), 337.

74. Robert Millikan, "The Diffusion of Science—The Natural Sciences," *Science Monthly*, September 1932, 205.

75. Rydell, "The Fan Dance of Science," 409.

76. Nelson B. Henry, ed., *46th Yearbook of the National Society for the Study of Education* (Chicago: University of Chicago Press, 1947), 62.

77. Larry Laudan, *Science and Hypothesis* (Dordrecht: D. Reidel, 1981), 1. See also Smith, *Fact and Feeling*, 13.

78. Millikan, "The Diffusion of Science" 204.

79. Peter Bergman, *Introduction to the Theory of Relativity* (New York: Prentice Hall, 1942), 28; Poor, "What Einstein Really Did," 529.

80. Percy Bridgeman, *The Logic of Modern Physics* (New York: Macmillan, 1927), 2–3 and vii–viii; idem, *The Nature of Physical Theory* (Princeton, NJ: Princeton University Press, 1936), 94.

81. Arthur Stanley Eddington, *The Mathematical Theory of Relativity* (Cambridge: Cambridge University Press, 1923), 3; idem, *The Nature of the Physical World*, 260. On behaviorism, see Kerry W. Buckley, *Mechanical Man: John Broadus Watson and the Beginnings of Behaviorism* (New York: Guildford Books, 1989); John

M. O'Donnell, *The Origins of Behaviorism: American Psychology, 1870–1920* (New York: New York University Press, 1985).

82. George D. Birkhoff, "The Origin, Nature and Influence of Relativity," *Science Monthly*, March 1924, 410.

83. David Starr Jordan, "The Nature of Sciosophy and Science," *Science Monthly*, February 1927, 150.

84. Schudson, *Discovering the News*, 7–8.

85. LeConte, "Agassiz and Evolution," 21; idem, "The Effect of the Theory of Evolution on Education," 123; John Merle Coulter, *Evolution, Heredity, and Eugenics* (Bloomington, IL: J. G. Coulter, 1916), 6–8 and 129–30; LaFollette, *Making Science*, 112.

86. Howard D. Roelofs, "In Search of Scientific Method," *American Scholar*, Summer 1940, 296, 301, and 304; Willard Waller, "Insight and Scientific Method," *American Journal of Sociology*, November 1934, 288.

87. "Einstein's Reality," *Time*, 16 March 1936, 74.

88. William Ritter, quoted in Watson Davis, "Science, Philosophy, Religion Find Ground for Common Front," *Science News Letter*, 21 September 1940, 190; Peter J. Kuznick, "Losing the World of Tomorrow: The Battle over the Presentation of Science at the 1939 New York World's Fair," *American Quarterly*, 46 (1994): 347.

89. Slossen, *Easy Lessons in Einstein*, 100, 107–8.

90. Percy Bridgeman, *The Logic of Modern Physics*, 2–3, vii–viii; idem, *The Nature of Physical Theory*, 94.

91. Robert Carmichael, *The Theory of Relativity* (New York: John Wiley & Sons, 1913), 64; James Jeans, *Mysterious Universe* (New York: Macmillan, 1930), 104; "Exile in Princeton," *Time*, 4 April 1938, 41.

92. Karl Pearson, *The Grammar of Science* (London: W. Scott, 1892), 31; Anton Reiser, *Albert Einstein* (New York: A. and C. Boni, 1930), quoted in Henry Hazlitt, "Einstein," *Nation*, 19 November 1930, 553; "Einstein's Reality," 74.

93. Poor, "What Einstein Really Did," 537; John T. Blankart, "Relativity or Interdependence," *Catholic World*, February 1921, 604; "God and Mathematics," *Nation*, 27 September 1933, 342.

94. F.S.C. Northrop, "The Theory of Relativity and the First Principles of Science," *Journal of Philosophy*, 2 August 1928, 422; Roelofs, 296; Edward S. Martin, "Einstein Gets Us Guessing," *Harper's*, April 1929, 654 ; "Einstein, Living Immortal," *Scientific American*, December 1930, 466; Albert Einstein, "Personal God Concept Causes Science-Religion Conflict," *Science News Letter*, 21 September 1940, 181; C. Hartley Grattan, "Why, Dr. Einstein!" *New Republic*, 9 March 1932, 95.

95. LaFollette, *Making Science*, 101.

96. This was true of articles with "scientist" in their titles as indexed by both *PCI* and *Readers' Guide*. These searches were conducted in January 2003.

97. LaFollette, *Making Science*, 101.

98. "Two to One—In Favor of Einstein," *Outlook*, 12 August 1925, 512.

99. James Gilbert, *Redeeming Culture: American Religion in an Age of Science* (Chicago: University of Chicago Press, 1997), 23–35.

100. James W. Cromer, *A Three-Dimensional Universe* (Los Angeles: 1924); idem, *A Side Note Criticism* (Los Angeles: 1925).

101. Tobey, *American Ideology*, 16–19.

102. E. E. Slossen, "How the Chemist Moves the World," *Science Monthly*, November 1922, 478.

103. William Hudgings, *Introduction to Einstein* (New York: Arrow Book Co., 1922), 7.

104. "The Einstein Theory at the Recent Meeting of the American Association for the Advancement of Science," *Scientific American*, March 1926, 188.
105. "One Score against Einstein," *Outlook*, 12 August 1925, 512–13.
106. Drawing at the top of Mortimer A. Morgan, "Are We About to Learn the Secrets of the Universe?" *Illustrated World*, September 1921, 26.
107. Photo, *Science Monthly*, April 1936, 381; "Time Stops in Its Tracks: An 'Atomic Clock' Proves Theory Einstein Voiced in 1905," *Newsweek*, 9 May 1938, 24.
108. Reuterdahl, *Einstein and the New Science*, 1; "Einstein Does It Again," *Review of Reviews*, March 1929, 128.
109. Slossen, *Easy Lessons in Einstein*, 107–8.
110. Hubert Kelley, "The Man from Space," *American Magazine*, September 1934, 128.
111. Sheldon, *Space, Time, and Relativity*, 12.
112. Archibald Henderson, "Einstein, a Picture of the Man," *Review of Reviews*, June 1929, 48.
113. Robert Merrill Bartlett, "Peace Must Be Waged," *Survey Graphic*, August 1935, 384.
114. Quoted in Grattan, "Why Dr. E!" 95.
115. "The Great Relativist Is Absolutely against War," *Christian Century*, 21 September 1931, 1085.
116. "Einstein: *Jewish Daily* Helped in Promotion by Herr Doktor," *Newsweek*, 27 January 1934, 31–32.
117. Edwin B. Wilson, "Space, Time and Gravitation," *Science Monthly*, March 1920, 218.
118. Priscilla Ingalls White, "Einstein Eclipsed," *Forum*, August 1930, xvii.
119. Grattan, "Why Dr. E!" 94–95.
120. "Einstein the Pacifist," *Christian Century*, 24 December 1930, 1581; "Pacifism and Relativity," *Christian Century*, 20 September 1933, 1163.
121. George DeBothezat, *Back to Newton* (New York: G. E. Stechert, 1936), 14.
122. M. K. Wisehart, "A Close Look at the World's Greatest Thinker," *American Magazine*, June 1930, 19.
123. William P. Montague, "Beyond Physics," *Saturday Review of Literature*, 23 March 1929, 800.
124. Ibid., 378.
125. See, for instance, William P. Montague, "Beyond Physics," *Saturday Review of Literature*, 23 March 1929, 800–801; William Morris Davis, "The Faith of a Reverent Scientist," *Science Monthly*, May 1934, 395–428.
126. Wilson, "Space, Time and Gravitation," 221.
127. Jeans, *Mysterious Universe*, 78–80, 84, and 158. On Millikan's work and its links with religion, see Kevles, *Physicists*, 179–80 and 239–42.
128. Albert Einstein, "Personal God Concept," *Science News Letter*, 21 September 1940, 181.
129. Frank Allen, *The Universe* (New York: Harcourt, Brace, 1931), 142.
130. Samuel H. Guggenheimer, *The Einstein Theory* (New York: Macmillan, 1925), 39.
131. Eddington, *Nature of the Physical World*, 309–10.
132. Hendrik Willem van Loon, "Short Cut to God," *New Republic*, 31 December 1930, 187.
133. "Einstein and Religion," 653–54; "Science—or Piffle," *Commonweal*, 19 November 1930, 60.
134. On public reaction to the Scopes trial, see Larson, *Summer for the Gods*; Edward Caudill, *Darwinism in the Press* (Hillsdale, NJ: L. Erlbaum Associates, 1989);

Ronald L. Numbers, "The Scopes Trial: History and Legend," in *Darwinism Comes to America*, 76–91.

135. This was true of articles with "science and religion" in their titles as indexed by both *PCI* and *Readers' Guide*. These searches were conducted in January 2003.

136. "Priests in the Religion of Science," *Review of Reviews*, January 1923, 120.

137. "Science and God: A German Dialogue," *Forum*, June 1930, 373.

138. Allen, *Only Yesterday*, 195–97.

139. Rydell, "The Fan Dance of Science," 401.

140. H. L. Mencken, "Nonsense as Science," *American Mercury*, December 1932, 509–10; Albert Einstein, *Relativity: The Special and General Theory*, trans. Robert W. Lawson (New York: H. Holt & Co., 1920), 55. On discussion of the deeper meanings of the fourth dimension, see Richard Erikson, *Consciousness, Life and the Fourth Dimension* (New York: A. A. Knopf, 1923); G. H. Paelian, *Relativity and Reality* (New York: Macoy, 1932); P. D. Ouspensky, *Tertium Organum* (New York: Knopf, 1922).

141. Henry Bedinger Mitchell, "The Theory of Relativity," *Theosophical Quarterly*, April 1929, 314.

142. Kevles, *Physicists*, 180; Max Talmey, *The Relativity Theory Simplified* (New York: Falcon Press, 1932), 102, 135, 147–48; Max Talmey, "Einstein's Theory and Rational Language," *Science Monthly*, September 1932, 254–57; Mencken, "Nonsense as Science," 509–10.

143. DeBothezat, *Back to Newton*, 10.

144. Arthur Lynch, *The Case against Einstein* (New York: Dodd, Mead, 1932), xi.

145. Carmichael et. al., *A Debate*, 124.

146. Allen, *Only Yesterday*, 197.

147. Crew, "The Exposition of Science," 228–29.

148. Quoted in Kevles, *Physicists*, 174.

149. Elsa Einstein, "You Lucky Americans," *Good Housekeeping*, August 1931, 34–35 and 187; Marchand, *Advertising the American Dream*, 2. Also see Ruth Cowan, *Social History of American Technology* (New York: Oxford University Press, 1997).

150. Erwin Schroedinger, "Einstein Explained," *World's Work*, June 1929, 146.

151. "Einstein's Latest Theory," *Popular Mechanics*, April 1929, 536.

152. This was true of articles with "applied science" in their titles as indexed by both *PCI* and *Readers' Guide*. These searches were conducted in January 2003.

153. Bertrand Russell, *ABC of Relativity* (New York: Harper & Brothers, 1925), 183.

154. J.W.N. Sullivan, *Three Men Discuss Relativity* (London: Collins, 1926), xii. See also idem, *Limitations of Science* (New York: Viking Press, 1934).

155. J. S. Ploughe, "Science Has Its Enigmas," *Christian Century*, 1 February 1933, 157.

156. "Chemistry: 300th U.S. Year; Synthetic Cancer and Rubber," *Newsweek*, 4 May 1935, 38.

157. Roslynn D. Haynes, "The Scientist in Literature," *Interdisciplinary Science Reviews* 14 (1989): 384.

Chapter 4   UFOs: In the Shadow of Science

1. "Lost Cause," *Nation*, 27 January 1969, 100; Edward Condon, "UFOs I Have Loved and Lost," *Bulletin of the Atomic Scientists*, December 1969, 6–8.

2. Victor Black, "The Flying Saucer Hoax," *American Mercury*, October 1952, 62.

3. "Those Flying Saucers: An Astronomer's Explanation," *Time*, 9 June 1952, 54.

4. Herschel W. Leibowitz, quoted in Robert Cowan, "Explanations of the First Kind," *Technology Review*, March/April 1979, 83.

5. "A Fresh Look at Flying Saucers," *Time*, 4 August 1967, 32.

6. J. Allen Hynek, with Barbara Ford, "Science Takes Another Look at UFOs," *Science Digest*, June 1973, 13.

7. This was true of articles with "pseudoscience" in their titles as indexed by *Readers' Guide* and *PCI*. These searches were conducted in January 2003.

8. "The Somethings," *Time*, 14 July 1947, 18.

9. "Saucer Season," *Newsweek*, 11 August 1952, 56. On early UFO sightings, see David Jacobs, *The UFO Controversy in America* (Bloomington: Indiana University Press, 1975), 35–63; Curtis Peebles, *Watch the Skies!* (Washington, DC: Smithsonian Institution Press, 1994), 18–36. On the 1952 flap, see Jacobs, *UFO Controversy*, 63–88; Peebles, *Watch the Skies*, 53–72.

10. This sequence of flaps was suggested in Stuart Nixon, "How to Report Things That Go Bump in the Night," *Quill*, July 1974, 20. On Carter's sighting, see Peebles, *Watch the Skies*, 204.

11. The 15 million figure was the result of a Gallup Poll taken in the mid-1970s. Results were reported in Ralph Blum, "Are UFOs for Real?" *Readers Digest*, June 1974, 92.

12. On the 1896/7 airship sightings, see Jacobs, *UFO Controversy*, 5–34; Daniel Cohen, *The Great Airship Mystery* (New York: Dodd, Mead, 1981). On Foo Fighters, see Jacobs, *The UFO Controversy*, 35–36; Peebles, *Watch the Skies*, 2; Cohen, *Airship Mystery*, 178–79.

13. "Visitors from Venus," *Time*, 9 January 1950, 49.

14. See, for instance, Michael S. Sherry, *The Rise of American Air Power: The Creation of Armageddon* (New Haven, CT: Yale University Press, 1987).

15. Willy Ley, "What Will 'Space People' Look Like?" *Science Digest*, February 1958, 62.

16. Norman J. Crum, "Flying Saucers and Book Selection," *Library Journal*, 1 October 1954, 1719; Edward J. Ruppelt, "Inside Story of the Saucers," *Science Digest*, April 1956, 35.

17. Jacobs, *UFO Controversy*, 41.

18. "Flying Saucers Are Old Stuff," *Popular Science*, August 1952, 147.

19. See Donald H. Menzel, *Flying Saucers* (Cambridge, MA: Harvard University Press, 1950). See also idem, "Saucers or Radar? . . . An Expert's Verdict," *Popular Science*, April 1953, 168–71, 268, and 296; idem, "A New Theory of Flying Saucers," *Science Digest*, September 1952, 11–16; "Biblical Flying Saucers," *Science News Letter*, 7 March 1953, 148; Robert Jastrow, "The Case for UFOs," *Science Digest*, November/December 1980, 85.

20. Jacques Vallee classified the late 1940s and early 1950s as the "American Period" of UFO sightings. He dated the "European Period" as beginning in the mid-1950s. See Jacques Vallee, *Challenge to Science* (Chicago: Regnery, 1966), 116–25.

21. On the initial Air Force investigation of UFOs, see Jacobs, *UFO Controversy*, 41–56.

22. On general links between postwar science and the federal government, see Kevles, *Physicists*, 324–66; Stuart Leslie, *The Cold War and American Science: The Military-Industrial-Academic Complex at MIT and Stanford* (New York: Columbia University Press, 1993); Alfred K. Mann, *For Better or for Worse: The Marriage of Science and Government in the United States* (New York: Columbia University Press, 2000), esp. 173–50. On the formation of the National Science Foundation, see Daniel Kevles, "The National Science Foundation and the Debate over Postwar Research Policy, 1942–1945: A Political Interpretation of 'Science—The Endless Frontier,'" *Isis* 68 (1977): 5–26; Daniel Kleinman, *Politics on the Endless Frontier: Postwar Research Policy in the United States* (Durham, NC: Duke University Press, 1995).

23. J. Allen Hynek, "Flying Saucers—Are They Real?" *Readers Digest*, March 1967, 63.

24. Jacques Vallee, "Unidentified Flying Objects," *Bulletin of the Atomic Scientists*, April 1973, 52; idem, *Invisible College* (New York: Dutton, 1975).

25. On the fight for congressional hearings, see Jacobs, *UFO Controversy*, 158–92; Peebles, *Watch the Skies*, 128–46.

26. For articles critical of the Condon study, see, for instance, John G. Fuller, "Flying Saucer Fiasco," *Look*, 14 May 1968, 59–63; Phillip M. Boffey, "UFO Project: Trouble on the Ground," *Science*, July 1968, 339–42. On the Condon Panel, see Jacobs, *UFO Controversy*, 225–63: Peebles, *Watch the Skies*, 169–95.

27. "We've Been Asked: Is There Something to UFO's After All?" *US News*, 20 February 1978, 56. For a survey of polls taken among astronomers, see Jerome Clark, ed., *The UFO Encyclopedia: The Phenomenon from the Beginning* (Detroit, MI: Omnigraphics, 1998), 28–30.

28. "End of 'The Search,'" *Newsweek*, 21 March 1955, 102–3.

29. Bruce Lewenstein, "Was There Really a Popular Science Boom?" *Science, Technology, and Human Values* 12 (1987): 29–41.

30. Black, "The Flying Saucer Hoax," 62.

31. This conclusion is based on a search of *Readers' Guide* conducted in January 2003.

32. Bruce Lewenstein, "The Meaning of 'Public Understanding of Science' in the United States after World War II," *Public Understanding of Science* 1 (1992): 45–68; idem, "Magazine Publishing and Popular Science after World War II," *American Journalism* 6 (1989): 218–34.

33. Robert C. Davis, *The Public Impact of Science in the Mass Media* (Ann Arbor: Survey Research Center, University of Michigan, 1958), 2–3 and 42.

34. Lewenstein, "Was There Really a Popular Science Boom?" 29–41.

35. "The Saucer-Eyed Dragons," *Time*, 17 April 1950, 52; Jacobs, *UFO Controversy*, 57.

36. Kenneth Arnold, "I *Did* See the Flying Disks!" *Fate*, Spring 1948, 4–12.

37. On Raymond Palmer, see Peebles, *Watch the Skies*, 3–6.

38. On Keyhoe and his work, see ibid., 56–62. On the formation of NICAP, see ibid., 132–57; Peebles, *Watch the Skies*, 109–18.

39. "On the Flying Saucer Trail," *American Magazine*, April 1954, 56; "Out-of-the-Blue Believers," *New Yorker*, 18 April 1959, 36.

40. For Vallee's survey of twenty-eight civilian UFO groups in the United States, see Vallee, *Challenge to Science*, 224–40. On the contactees, see Jacobs, *UFO Controversy*, 108–31; Peebles, *Watch the Skies*, 93–108. On the roots of contactee lore in earlier theosophist ideas, see David Stupple, "Mahatmas and Space Brothers: The Ideologies of Alleged Contact with Extraterrestrials," *Journal of American Culture* 7 (1984): 131–39.

41. On the controversy over Velikovsky's work, see Henry H. Bauer, *Beyond Velikovsky: The History of a Public Controversy* (Urbana: University of Illinois Press, 1985).

42. Carl Sagan, introduction to *UFOs: A Scientific Debate*, ed. Carl Sagan and Thornton Page (Ithaca, NY: Cornell University Press, 1972), xiii.

43. Thornton Page, "Student Skeptics Study UFO's," *Science*, 15 December 1967, 1397.

44. Billy Ingram, "Short History of the Tabloids: 50+ Years of the *National Enquirer*," http://www.tvparty.com/tabloids.html (18 May 2006).

45. On the counterculture in the United States during the 1960s and 1970s, see Peter Braunstein and Michael William Doyle, eds., *Imagine Nation: The American Counterculture of the 1960s and 1970s* (London: Routledge, 2001); Alice Echols, *Shaky*

Ground: The '60s and Its Aftershocks (New York: Columbia University Press, 2002); Stewart Burns, *Social Movements of the 1960s* (Boston: Twayne Publishers, 1990); Anthony M. Casale and Philip Lerman, *Where Have All the Flowers Gone? The Fall and Rise of the Woodstock Generation* (Kansas City, MO: Andrews and McMeel, 1989); Robert V. Daniels, *The Fourth Revolution: Transformations in American Society from the Sixties to the Present* (London: Routledge, 2006); Mark Oppenheimer, *Knocking on Heaven's Door: American Religion in the Age of Counterculture* (New Haven, CT: Yale University Press, 2003); Irwin Unger and Debi Unger, eds., *The Times Were a Changin': The Sixties Reader* (New York: Three Rivers Press, 1998).

46. See, for instance, Joel Hagen, *An Entangled Bank: The Origins of Ecosystem Ecology* (New Brunswick, NJ: Rutgers University Press, 1992), esp. 100–145. On Rachel Carson and *Silent Spring*, see Zuoyue Wang, "Responding to 'Silent Spring': Scientists, Popular Science Communication, and Environmental Policy in the Kennedy Years," *Science Communication* 19 (1997): 141–63.

47. Quoted in Kevles, *Physicists*, 401.

48. Dorothy Nelkin, *Selling Science* (New York: W. H. Freeman, 1987), 91–92.

49 Amitai Etzoni and Clyde Nunn, "The Public Appreciation of Science in Contemporary America," in *Science and Its Publics*, ed. William Holton and Gerald Blanpied (Dordrecht: Reidel, 1976), 232.

50. Kevles, *Physicists*, 415.

51. "The UFO Clans Gather," *Time*, 3 November 1975, 64.

52. On the abductee movement, see Peebles, *Watch the Skies*, 225–41.

53. Burnham, *How Science Lost*, 252.

54. Ross, *"Scientist,"* 83.

55. This was true of articles with "scientist" in their titles as indexed by both *PCI* and *Readers' Guide*. These searches were conducted in January 2003.

56. Don Berliner, "Study Challenges Sky Mirages as a Sighting Source," *Science Digest*, August 1978, 31–32.

57. Desmond Leslie and George Adamski, *Flying Saucers Have Landed* (London: W. Laurie, 1953), 48 and 67–68.

58. Numbers, *Creationists*, 60 and 184ff.

59. Mead and Metraux, "Image of the Scientist," 387; Davis, *The Public Impact of Science*, 190.

60. Mead and Metraux, "Image of the Scientist," 387; quoted in Haynes, "The Scientist in Literature," 384.

61. LaFollette, *Making Science*, p. 90.

62. Vallee, "Unidentified Flying Objects," 49.

63. Nelkin, *Selling Science*, 65.

64. "More Saucers," *Time*, 3 March 1952, 92 and 94; "Ufology," *Newsweek*, 5 August 1963, 44.

65. "UFO's for Real?" *Newsweek*, 10 October 1966, 70.

66 Philip Wylie, "UFOs: The Sense and Nonsense," *Popular Science*, March 1967, 79; "Seeing Things," *Newsweek*, 18 November 1957, 41–44. The prominence of hoaxes in most media discussion went far beyond that warranted by the less than 2 percent of sightings identified as frauds by the Air Force .On discussion of the supposed frequency of hoaxes, see Clare Boothe Luce, "Without Portfolio," *McCall's*, July 1967, 32; "ATIC Begins Study of Saucer Reports," *Aviation Week*, 19 October 1953, 18; "A Hard Look at 'Flying Saucers,'" *US News*, 11 April 1966, 15.

67. Carl Sagan, "Unidentified Flying Objects," *Bulletin of the Atomic Scientists*, June 1967, 43. The same claim is made in "No Evidence for Saucers," *Science News Letter*, 16 November 1957, 307; "Finds 'Saucers' Exist Solely in Imagination,"

Science Digest, January 1955, 24; Daniel Cohen, "Should We Be Serious About UFO's?" *Science Digest*, June 1965, 44.

68. Hynek, quoted in "Saucer Diehards," *Time*, 28 June 1971, 49. One survey of 2,611 members of the American Astronomical Society was reported in Patrick Huyghe, "Scientists Who Have Seen UFOs," *Science Digest*, November 1981, 94. Other surveys are summarized in Jerome Clark, *The UFO Encyclopedia* (Detroit, MI: Apogee Books, 1990), 29.

69. "A Fresh Look at Flying Saucers," *Time*, 4 August 1967, 33.

70. "UFO Consensus," *Science*, 8 December 1967, 1266; Donald Menzel, "Case against UFO's," *Physics Today*, June 1976, 15. On the meteorite story, see "A Fresh Look at Flying Saucers," 32; David Hufford, "Humanoids and Anomalous Lights: Taxonomic and Epistemological Problems," *Fabula* 18 (1977): 239; Brendan I. Koerner, "Lessons of the UFOs," *US News*, 13 July 1998, 52; John C. Munday, Jr., "On the UFOs," *Bulletin of the Atomic Scientists*, December 1967, 41; Edith Kermit Roosevelt, "Those Flying Saucers Again!" *American Mercury*, September 1956, 156; J. Allen Hynek, "UFO's Merit Scientific Study," *Science*, 21 October 1966, 329; David Kestenbaum, "Panel Says Some UFO Reports Worthy of Study," *Science*, 3 July 1998, 21.

71. Huyghe, "Scientists Who Have Seen UFOs," 119.

72. J. Allen Hynek, *UFO Experience* (Chicago: H. Regnery Co., 1972), 237. Vallee repeated this charge. See Vallee, "Unidentified Flying Objects," 49.

73. Curtis Fuller, "I See by the Papers," *Fate*, November 1969, 7.

74. "Dinner Time," *Time*, 18 November 1957, 28.

75. John Fuller, "Did Flying Saucers Cause the Blackout?" *Popular Mechanics*, November 1966, 241.

76. Paul O'Neil, "A Well-Witnessed 'Invasion'—by Something," *Life*, 1 April 1966, 29.

77. "UFO's for Real?" 70; Page, "Student Skeptics Study UFO's," 1397; D.C., "Whatever Happened to Flying Saucers?" *Science Digest*, September 1963, 64.

78. These numbers are based on a collection of 245 individuals cited between 1947 and 2000.

79. LaFollette, *Making Science*, 117–22.

80. "Reporter Edward Condon," *Saturday Review*, 1 February 1969, 55–56.

81. Karlheinz Steinmuller, "Science Fiction and Science in the Twentieth Century," *Science in the Twentieth Century*, 359.

82. "'UFO's'—They're Back," 60.

83. "Plenty Going on in the Skies," *US News*, 1 January 1954, 28; "What's Going on in the Skies?" *US News*, 8 August 1952, 13; "If You're Seeing Things in the Sky Air Force Can Account for Most of Them, But . . . ," *US News*, 15 November 1957, 126; Roosevelt, "Those Flying Saucers Again!" 153.

84. Kevles, *Physicists*, 401–3.

85. "Closing the Blue Book," *Time*, 26 December 1969, 28.

86. Huyghe, "Scientists Who Have Seen UFOs," 119.

87. "Closing the Blue Book," 28.

88. Wylie, "UFOs: The Sense and Nonsense," 76; Hynek, "Flying Saucers—Are They Real?" 65.

89. Steven Shapin has claimed that institutions have come to eclipse individuals as guarantors of truthfulness in the modern world. See Steven Shapin, *A Social History of Truth* (Chicago: University of Chicago Press, 1994), 412–13.

90. John G. Fuller, "Trade Winds," *Saturday Review*, 22 January 1966, 14.

91. Jim Wilson, "Six Unexplainable Sightings," *Popular Mechanics*, July 1998, 67.

92. Blum, "Are UFOs for Real?" 90.

93. Ted Bloecher, "The Case for Flying Saucers," *Saturday Review*, 27 August 1955, 21.
94. Frank Scully, *Behind the Flying Saucers* (New York: Holt, 1950), xi.
95. Haynes, "The Scientist in Literature," 384.
96. O'Neil, "A Well-Witnessed 'Invasion'—by Something," 29. On the Oppenheimer and Condon cases, see Kevles, *Physicists*, 378–80; Rachel Halloway, *In the Matter of J. Robert Oppenheimer* (Westport, CT: Praeger, 1993).
97. John Mulholland, "Magicians Scoff at Flying Saucers," *Popular Science*, September 1952, 97.
98. "UFO Consensus," 1265.
99. David Cort, "Saucery and Flying Saucers," *Nation*, November 1959, 340.
100. Ad for Gumout, *Popular Mechanics*, September 1997, 42–43; James Mullaney, "UFOlogy," *Science Digest*, July 1977, 28; "Flying Saucers and Science," *American Mercury*, July 1957, 125.
101. "Martians over France," *Time*, 25 October 1954, 71.
102. William L. Moore, *The Philadelphia Experiment* (New York: Grosset & Dunlap, 1979). A quick search on the Internet will reveal a wide variety of sites discussing "Beefield-Brown" generators.
103. On the abduction phenomenon, see John Fuller, *Interrupted Journey* (New York: Dial Press, 1966); Coral Lorenzen and James Lorenzen, *Abducted: Confrontations with Beings from Outer Space* (New York: Berkeley, 1977); Budd Hopkins, *Missing Time* (New York: Marek, 1981); idem, *Intruders* (New York: Ballantine Books, 1987); Whitley Strieber, *Communion* (New York: William Morrow, 1987); David M. Jacobs, *Secret Life* (New York: Simon & Schuster, 1993); John E. Mack, *Abduction* (New York: Scribner's, 1994).
104. Davis, *The Public Impact of Science*, 186. On alternative technology, see Andrew Kirk, "'Machines of Loving Grace': Alternative Technology, Environment, and Counter Culture," in *Imagine Nation: The American Counter Culture of the 1960s and 1970s*, ed. Peter Braunstein and Michael William Doyle (New York: Routledge, 2002).
105. This was true of articles with "science" and "nature" in their texts as indexed by *LexisNexis*, http://web.lexis-nexis.com/universe. These searches were conducted in January 2003.
106. Cowan, "Explanations of the First Kind," 83.
107. "A Hard Look at 'Flying Saucers,'" *US News*, 11 April 1966, 15.
108. Hynek, *UFO Experience*, 237; "A Hard Look at 'Flying Saucers,'" 15; Huyghe, "Scientists Who Have Seen UFOs," 119.
109. Condon, "UFOs I Have Loved and Lost," 8.
110. This was true of articles with "scientific method" in their titles as indexed by both *PCI* and *Readers' Guide*. These searches were conducted in January 2003.
111. See, for instance, Stephenie G. Edgerton, "Is There a Scientific Method?" *History of Education Quarterly*, Winter 1969, 492–95; A. Cornelius Benjamin, "Is There a Scientific Method?" *Journal of Higher of Education*, May 1956, 233–38; Joseph Turner, "Is There a Scientific Method?" *Science*, 6 September 1957, 431.
112. Helen P. Libel, "History and the Limitations of Scientific Method," *University of Toronto Quarterly*, October 1964, 15–16.
113. Rudolph, *Scientists in the Classroom*.
114. Davis, *The Public Impact of Science*, 203.
115. Roland Gelatt, "In a Saucer from Venus," *Saturday Review of Literature*, 23 September 1950, 20.
116. Waldemar Kaempffert, "Remember the Flying Saucers?" *Science Digest*, October 1947, 70.
117. Philip Wylie, "Of Stress and Saucers," *Saturday Review*, 8 August 1959, 17.

118. Constance Holden, "AF Bestows on National Archives a Trove for UFOlogists," *Science*, 20 August 1976, 663.
119. Basalla, "Pop Science"; "Cause of 'Flying Saucers,'" *Science News Letter*, 9 August 1952, 82; "'UFO's'—They're Back," 60. On invocation of technical training, see Howard Margolis, "The UFO Phenomenon," *Bulletin of the Atomic Scientists*, June 1967, 40; Hynek with Ford, "Science Takes Another Look," 12; Roosevelt, "Those Flying Saucers Again!" 154; Hynek, "Flying Saucers—Are They Real?" 62; "More Saucers," *Time*, 3 March 1952, 92; "Ufology," *Newsweek*, 5 August 1963, 44. For final quote, see "Tracking UFOs Scientifically," *Science Digest*, May 1976, 49.
120. Philip M. Buffey, "UFO Study: Condon Group Finds No Evidence of Visits from Outer Space," *Science*, 17 January 1969, 262.
121. Ronald Tobey, "Epiphany and Conspiracy: The UFO Controversy," *Reviews in American History*, March 1976, 130.
122. Davis, *The Public Impact of Science*, 184.
123. James Willwerth, "The Man from Outer Space," *Time*, 25 April 1994, 75.
124. On Capra's television specials, see Gilbert, *Redeeming Culture*, 199–223. On religious metaphor in science fiction, including *The Day the Earth Stood Still*, see ibid., 225–51.
125. Davis, *The Public Impact of Science*, 193.
126. Mullaney, "UFOlogy," 31–32; Vallee, "Unidentified Flying Objects," 51–52.
127. See, for instance, Keith Thompson, *Angels and Aliens* (Reading, MA: Addison-Wesley, 1991).
128. George Adamski, *Inside the Space Ships* (New York: Abelard-Schuman, 1955), 182; Erich von Daniken, *Chariots of the Gods*, trans. Michael Heron (New York: Putnam, 1968); R. L. Dione, *God Drives a Flying Saucer* (New York: Bantam Books, 1969). On Meade Layne and the Borderland Sciences Research Associates, see Leslie and Adamski, *Flying Saucers Have Landed*, 125–32; Clark, "The Extraterrestrial Hypothesis," 125–26.
129. "Letters to the Editor," *Look*, 1 December 1953, 14.
130. "Heavenly Bogeys," *Time*, 2 September 1966, 81.
131. Robert H. Wood, "Where Are the Flying Saucers," *Aviation Week*, 25 June 1951, 74.
132. On the use of "cultists," see "Saucer Blue Book," *Time*, 7 November 1955, 52; Carl Sagan, quoted in Cohen, "UFOs—What a New Investigation May Reveal," 62; "Waiting for the Little Men," *Newsweek*, 28 March 1955, 64; Condon, "UFOs I Have Loved and Lost," 7. On the invocation of "believers" and other general religious language to describe UFO advocates and their convictions, see "Spacemen Don't Fly Saucers," *Senior Scholastic*, 12 January 1970, 2–3; "Lost Cause," 100; "Condon Study Rebuts UFOs; Critics Offer Own Version," *Physics Today*, March 1967, 71; Edward Ziegler, "What's Behind Our UFO Mania?" *Readers Digest*, August 1987, 113–17; "Those Flying Saucers," 54–56; "In This Issue," *Christian Century*, 22 February 1978, 178; "The UFO Clans Gather," 64; James S. Gordon, "The UFO Experience," *Atlantic Monthly*, August 1991, 82–86 and 88–92; "Saucer's End," *Time*, 17 January 1969, 44–45; "UFO's for Real?" 70; Daniel Cohen, "Should We Be Serious about UFO's?" *Science Digest*, June 1965, 41–44; "Tracking UFOs Scientifically," 42–49. On "sci-fi religion," see Stephen J. Hodges, "www.masssuicide.com," *US News*, 7 April 1997, 26. On the invocation of unfulfilled religious need, see Sagan, "Unidentified Flying Objects," 44.
133. Irving Hexham, "UFOlogy and Christianity," *Christianity Today*, 10 March 1978, 55.

134. This is based on searches of *Readers' Guide* for articles with "science and religion" and "science and technology" in their titles that were conducted in January 2003. The former was eclipsed by the latter sometime after 1940.
135. Vallee, *Invisible College*, 209.
136. Burnham, *How Science Lost*, 261–62.
137. Jacques Vallee, *Forbidden Science: Journals, 1957–1969* (Berkeley, CA: North Atlantic Books, 1992).
138. Mencken, "Nonsense as Science," 509.
139. John Pfeiffer, "Scientists Combine to Combat Pseudoscience," *Psychology Today*, November 1977, 38.
140. On pseudo-science, see Seymour H. Mauskopf, "Marginal Science," in *Companion to the History of Modern Science*, 869–85; R. Wallace, "Science and Pseudo-Science," *Social Science Information* 24 (1985): 585–601; Harry Collins and Trevor Pinch, *Frames of Meaning: The Social Construction of Extraordinary Science* (London: Routledge, 1982); Thomas Leahey and Grace Leahey, *Psychology's Occult Doubles: Psychology and the Problem of Pseudoscience* (Chicago: Nelson-Hall, 1983); Rachel Laudan, ed., *The Demarcation between Science and Pseudo-Science* (Blacksburg, VA: Center for the Study of Science in Society, Virginia Polytechnic Institute and State University, 1983); Roy Wallis, ed., *On the Margins of Science* (Keele: University of Keele, 1979).
141. J. Allen Hynek, "UFOs and the Numbers Game," *Natural History*, March 1968, 24.
142. Sagan, introduction to *UFO's: A Scientific Debate*, xiii.
143. Charles Fort, *The Book of the Damned* (New York: Ace Books, 1919), 18–28.
144. Tiffany Thayer, introduction to *The Books of Charles Fort*, by Charles Fort (New York: H. Holt & Co., 1941), xxiii–xxiv.
145. John A. Keel, *The Mothman Prophecies* (New York: Saturday Review Press, 1975).
146. Jule Eisenbud, "Why You Refuse to Believe," *Fate*, April 1969, 90.
147. On Fort's early novels, see Damon Knight, *Charles Fort: Prophet of the Unexplained* (London: Victor Gollancz, 1971), 51–62.
148. Vallee, *Invisible College*, 108–11. David Jacobs, who himself became involved in UFO research, also claimed that UFO encounters became increasingly weird from the 1950s to the 1970s. See Jacobs, "UFOs and the Search for Scientific Legitimacy," in *The Occult in America: New Historical Perspectives*, ed. Howard Kerr and Charles L. Crow (Urbana: University of Illinois Press, 1983), 218–31.
149. Leahey and Leahey, *Psychology's Occult Doubles*, 242.
150. Leslie and Adamski, *Flying Saucers Have Landed*, 71.
151. *Fate*, January 1971, 2.
152. Knight, *Charles Fort*, 2.
153. Andrew Ross, *Strange Weather* (New York: Verso, 1991), 15–74.
154. On Roswell, see Charles Berlitz and William Moore, *The Roswell Incident* (New York: Grosset & Dunlap, 1980); Kevin Randle and Donald Schmitt, *The Truth about the UFO Crash at Roswell* (New York: Avon Books, 1994); Philip J. Corso, *The Day after Roswell* (New York: Pocket Books, 1997); Kal Karff, *The Roswell UFO Crash* (Amherst, NY: Prometheus Books, 1997); Benson Saler, Charles A. Ziegler, and Charles B. Moore, *UFO Crash at Roswell: The Genesis of a Modern Myth* (Washington, DC: Smithsonian Institution Press, 1997).
155. On Heaven's Gate, see Brad Stieger, *Inside Heaven's Gate* (New York: Signet, 1997); Kenneth L. Woodward, "Christ and Comets," *Newsweek*, 7 April 1997, 40–43.
156. See, for instance, John E. Mack, *Abduction: Human Encounters with Aliens* (New York: Scribner's, 1994); David M. Jacobs, *Secret Life: Firsthand Accounts of UFO*

Abductions (New York: Simon & Schuster, 1992); Whitley Strieber, *Communion* (New York: Beech Tree Books, 1987).

157. On the Raelians, see Susan J. Palmer, *Aliens Adored: Rael's UFO Religion* (New Brunswick, NJ: Rutgers University Press, 2004). On Raelian claims about alien interference in evolution, see Robert T. Pennock, *Tower of Babel: The Evidence against the New Creationism* (Cambridge, MA: MIT Press, 1999), 234–35.

158. On Daniken's ideas, see, for instance, Erich von Daniken, *Chariots of the God?* (New York: Putnam, 1969).

Chapter 5   Intelligent Design: The Evolution of Science Talk

1. The appeal to objectivity was made on http://www.kansasscience2005.com/ (10 April 2006).

2. *Kansas State Education Science Standards*, draft 2D (2005), ix.

3. Robert Dennison, "Review of 'Proposed Revisions to Kansas Science Standards Draft 1,'" http://www.ksde.org/outcomes/sciencereviewrobertdennison.pdf (1 April 2006).

4. Keith Miller, quoted in Debora MacKenzie, "A Battle for Science's Soul," *New Scientist*, 9 July 2005, 8.

5. See Numbers, *The Creationists*; idem, "Introduction: Darwinism, Creationism, and Intelligent Design," in *Darwinism Comes to America*, 1–23.

6. See, for example, Niles Eldredge, review of *Evolution: A Theory in Crisis*, *Quarterly Review of Biology* 61 (1986): 541–42; Mark Ridley, "More Darwinian Detractors," *Nature* 318 (1985): 124–25; Philip T. Spieth, review of *Evolution: A Theory in Crisis*, *Zygon* 22 (1987): 252–57.

7. John G. West, Jr., "The Regeneration of Science and Culture," in *Signs of Intelligence*, ed. William A. Dembski and James M. Kushiner (Grand Rapids, MI: Brazos Press, 2001), 61.

8. "About Discovery: How Discovery Institute Functions," http://www.discovery.org/aboutfunctions.php (12 March 2006).

9. See, for instance, "U. of Kansas Chancellor Assails 'Anti-Science' Forces in His State," *Chronicle of Higher Education*, 7 October 2005, A12; Barbara Forrest and Paul R. Gross, *Creationism's Trojan Horse: The Wedge of Intelligent Design* (Oxford: Oxford University Press, 2004), 9; Nicholas Wapshott, "A New Age of Unreason," *New Statesman*, 17 October 2005, 36–37; Barbara Forrest, quoted in Larry Witham, "Intelligent Design on Trial," *Christian Century*, 29 November 2005, 9.

10. Forrest and Gross, *Trojan Horse*. There are two chapters with this title. See pp. 147–214.

11. Michael Friedlander, "Letters to the Editor," *Chronicle of Higher Education*, 14 October 2005, B13.

12. Michael J. Behe, "The God of Science: The Case for Intelligent Design," *The Weekly Standard*, 7 June 1999, 36.

13. Allan H. Rysking, "Darwinist Ideologues Are on the Run," *Human Events*, 30 January 2006, 7.

14. Forrest and Gross, *Trojan Horse*, 69.

15. Kendrick Frazier, "Court Decision in Dover Case a Victory for Good Science," *Skeptical Inquirer*, March/April 2006, 4.

16. P. David Hornik, "Deny the Designer, Save 'Science,'" *American Spectator Online*, 23 January 2006, http://www.spectator.org/dsp_article.asp?art_id=9307 (4 April 2006).

17. Michael J. Behe, *Darwin's Black Box* (New York: Free Press, 1996), 177–78 Behe's search was also reported in Dan Peterson, "The Little Engine That Could . . . Undo Darwinism," *American Spectator*, June 2005, 37.

18. Friedlander, "Letters to the Editor," B13.
19. Tom Junod, "The Case for Intelligent Design," *Esquire*, November 2005, 186.
20. "AGU: President Confuses Science and Belief, Puts Schoolchildren at Risk," *Skeptical Inquirer*, November/December 2005, 45.
21. Diane Cole, "Learn to Think Like a Scientist," *US News & World Report*, 26 December 2005, 62; Margaret Talbot, "Darwin in the Dock: Intelligent Design Has Its Day in Court," *New Yorker*, 5 December 2005, 68; See also David Morrison, "Only a Theory? Framing the Evolution/Creation Issue," *Skeptical Inquirer*, November/December 2005, 29.
22. Michael Ruse, *Darwinism Defended: A Guide to the Evolution Controversies* (Reading, MA: Addison-Wesley, 1982), 58. William Dembski mentions this claim by Ruse in "What Intelligent Design Is Not," in *Signs of Intelligence*, 14.
23. Lawrence Krauss, "Mind Your Language: Misusing the Word 'Theory' Plays into the Hands of Creationists," *New Scientist*, 3 December 2005, 23.
24. Michael Blanco, "Letters to the Editor," *Chronicle of Higher Education*, 14 October, 2005, B13.
25. Quoted in William Lee Adams, "Other Schools of Thought: The Teaching of Evolution Continues to Polarize Communities," *Newsweek*, 28 November 2005, 57.
26. Behe, "The God of Science," 57.
27. Dan Peterson, "What's the Big Deal about Intelligent Design?" *American Spectator*, December 2005/January 2006, 34.
28. Gene Edward Veith, Jr., "Science's New Heresy Trial," *World Magazine*, 19 February 2005, http://www.worldmag.com/articles/10344 (2 April 2006); Phillip E. Johnson, "The Intelligent Design Movement," in *Signs of Intelligence*, 31.
29. "U. of Kansas Chancellor Assails 'Anti-Science' Forces in His State," *Chronicles of Higher Education*, 7 October 2005, A12.
30. Christopher Bonanos, "Darwin," *New York Magazine*, 5 December 2005, 94. See also Ben McGrath, "Darwin in Manhattan," *New Yorker*, 21 November 2005, 42.
31. See Hornik, "Deny the Designer"; Stafford Betty, "Intelligent Design Theory Belongs in the Science Classroom," *National Catholic Reporter*, 21 October 2005, 23.
32. Jonathan Alter, "Monkey See, Monkey Do," *Newsweek*, 15 August 2005, 27; "AGU: President Confuses Science and Belief," 45; Junod, "The Case for Intelligent Design," 187.
33. Will Durst, "Unintelligent Design," *Progressive*, October 2005, 46.
34. Lawrence Lerner, "What Should We Think about Americans' Beliefs Regarding Evolution?" *Skeptical Inquirer*, November/December 2005, 60.
35. "Fake ID: Pa. Lawsuit Should Expose Creationist Foray into Public Schools," *Church & State*, October 2005, 13.
36. See, for instance, "Judge Rules Intelligent Design Is Not Science," *Christian Century*, 10 January 2006, 13.
37. Niall Shanks, *God, the Devil, and Darwin* (New York: Oxford University Press, 2004), 6; William Safire, "Neo-Creo," *New York Times Magazine*, 21 August 2005, 16; Forrest and Gross, *Trojan Horse*, 6; Durst, "Unintelligent Design," 46.
38. Betty, "Intelligent Design Theory Belongs in the Science Classroom," 23.
39. Cornelius G. Hunter, "Why Evolution Fails the Test of Science," in *Uncommon Dissent*, ed. William A. Dembski (Wilmington, DE: ISI Books, 2004), 214.
40. Michael Ruse, "Science under Siege," *Christian Century*, 15 November 2005, 31.
41. Matt Taibbi, "Darwinian Warfare," *Rolling Stone*, 3 November 2005, 46.
42. Shanks, *God, the Devil, and Darwin*, 14.

43. John Derbyshire, "Teaching Science," *National Review*, 30 August 2005, http://article.nationalreview.com/?q=YjFiZmIxODNhZGYwMDViYzM0ZjA50Tg4MTJmMGZhYmE= (5 April 2006).

44. Quoted in Claudia Wallis, "The Evolution Wars," *Time*, 15 August 2005, 28.

45. Alter, "Monkey See, Monkey Do," 27; Robert George Sprackland, "A Scientist Tells Why 'Intelligent Design' Is NOT Science," *Educational Digest*, January 2006, 33.

46. Jerry Adler, "Doubting Darwin," *Newsweek*, 7 February 2005, 46; Tom Bethell, "Politically Incorrect Science," *American Spectator*, November 2005, 46.

47. "Harris Poll Explores Beliefs about Evolution, Creationism, and Intelligent Design," *Skeptical Inquirer*, November/December 2005, 56; Sprackland, "A Scientist Tells Why," 30.

48. Chris Mooney and Matthew Nisbet, "Undoing Darwin," *Columbia Journalism Review*, September/October 2005, 30–39.

49. Behe marked technical paragraphs throughout the text with a special symbol and also included one specialized chapter. Dembski only did the latter. On their warnings about technical content, see Behe, *Darwin's Black Box*, xii. William Dembski, *Intelligent Design: The Bridge Between Science & Theology* (Downer's Grove, IL: InterVarsity Press, 1999), 15.

50. Quoted in Lawrence Krauss, "Mind Your Language: Misusing the Word 'Theory' Plays into the Hands of Creationists," *New Scientist*, 3 December 2005, 23.

51. Lerner, "What Should We Think?" 60.

52. Lerner, "What Should We Think?" 60; Steve Pinker, quoted in David Van Biema, "Can You Believe in God and Evolution?" *Time*, 15 August 2005, 34.

53. Forrest and Gross, *Trojan Horse*, 8 and 13.

54. Junod, "The Case for Intelligent Design," 187.

55. Hornik, "Deny the Designer."

56. Nancy R. Pearcey, "Darwin Meets the Berenstain Bears: Evolution as a Total Worldview," in *Uncommon Dissent*, 72–73.

57. Sprackland, "A Scientist Tells Why," 30.

58. David Morrison, "Only a Theory? Framing the Evolution/Creation Issue," *Skeptical Inquirer*, November/December 2005, 40.

59. Keith B. Miller, "The Controversy over the Kansas Science Standards," http://www.wheaton.edu/ACG/essays/miller1.html (4 March 2006).

60. Eric Cornell, "What Was God Thinking? Science Can't Tell," *Time*, 14 November 2005, 100.

61. John Boler, "What's Scientific about It?" *Commonweal*, 4 November 2005, 9.

62. Ruse, "Science under Siege," 32.

63. Quote in David Van Biema, "Can You Believe in God and Evolution?" *Time*, 15 August 2005, 34.

64. Sprackland, 31.

65. Behe, *Darwin's Black Box*, 30.

66. Bethell, "Politically Incorrect Science," 44–45.

67. Edward Sisson, "Teaching the Flaws in Neo-Darwinism," in *Uncommon Dissent*, 85.

68. William Dembski, *The Design Revolution* (Downer's Grove, IL: InterVarsity Press, 2004), 302.

69. Dembski, *Intelligent Design*, 247. On many other occasions, Dembski presents as models for the inclusion of intelligent design into the scientific enterprise a variety of design-based sciences (forensic science, cryptography, archeology, artificial intelligence, and the search for extraterrestrial intelligence) that sometimes *do* deal with motivation, purpose, character, and the like. See p. 127. But he does not extend such powers to ID.

70. Behe, *Darwin's Black Box*, 5.
71. Behe confined larger issues of this type to a final chapter of *Darwin's Black Box*. See Behe, *Darwin's Black Box*, 235–53. Dembski has tended to touch on them throughout his works.
72. Dembski, *Intelligent Design*, 229.
73. Nancy Pearcey, "Design and the Discriminating Public," in *Signs of Intelligence*, 44–45.
74. James Barham, "Why I Am Not a Darwinist," in *Uncommon Dissent*, 178; John G. West, Jr., "The Regeneration of Science and Culture," in *Signs of Intelligence*, 64.
75. Dembski, *Intelligent Design*, 100.
76. West, "The Regeneration of Science and Culture," 69.
77. Peterson, "The Little Engine That Could," 43.
78. Dembski, *Intelligent Design*, 114.
79. David Klinghoffer, "It's God or Darwin," *National Review Online*, 21 December 2005, http://article.nationalreview.com/?q=MjMzZDU2NmYzYWExMGNhMzgzODU3NTYzNGIzNDVhZTQ= (3 April 2006).
80. Behe, *Darwin's Black Box*, 231.
81. Dembski, *Intelligent Design*, 151, 233, and 121.
82. Pearcey, "Design and the Discriminating Public," 49.
83. John Wilson, foreword, in *Uncommon Dissent*, xiii.
84. Dembski, *Intelligent Design*, 206.
85. Pearcey, "Design and the Discriminating Public," 49.
86. Robert C. Koons, "The Check Is in the Mail: Why Darwinism Fails to Inspire Confidence," in *Signs of Intelligence*, 7. Dembski also attributes public resistance to evolution, as indicated in polling, to an intuitive sense that evolutionary models cannot fully explain the history of life. See Dembski, "What Intelligent Design Is Not," 14.
87. Ray Miller, "Science? Fiction?" *Texas Monthly*, June 2005, 16.
88. Behe, *Darwin's Black Box*, 201–3. It is interesting to note that ID grew up over almost precisely the same period of time as nanotechnology, which was coined as a public term in 1986 and which in some of its forms stresses the manufacture of molecular machines. See K. Eric Drexler, *Engines of Creation* (Garden City, NY: Anchor Press/Doubleday, 1986). To a much lesser extent than ID, Drexler's vision of nanotechnology, which has often made an appearance in science fiction, sometimes also borrowed religious terminology, as in the title of his book.
89. Dembski, *The Design Revolution*, 312; idem, *Intelligent Design*, 108.
90. Kenneth Lundgren, "Evolution Debate Engages Readers on Both Sides of the Argument," *Electronic Engineering Times*, 20 February 2006, 26.
91. Dembski, "What Intelligent Design Is Not," 14–15; Phillip E. Johnson, "Evolution as Dogma: The Establishment of Naturalism," in *Uncommon Dissent*, 37.
92. Dembski, *Intelligent Design*, 118.
93. Sisson, "Teaching the Flaws in Neo-Darwinism," 93.
94. Forrest and Gross, *Trojan Horse*, 8.
95. Charles P. Pierce, "Greetings from Idiot America," *Esquire*, November 2005, 181.
96. Quoted in Mike Lafferty, "Intelligent Design: Like Ohio, Kansas Redefined 'Science,'" *Columbus (OH) Dispatch*, 17 November 2005, 5D.
97. Dembski, *Intelligent Design*, 227.
98. Ibid., 53. The second claim was a quotation from John Calvert.
99. Dembski, *Design Revolution*, 21; idem, "Introduction: The Myths of Darwinism," in *Uncommon Dissent*, xxiii. The first quote was Dembski's paraphrase of a claim made by C. S. Lewis.

100. Phillip E. Johnson, "The Intelligent Design Movement," in *Signs of Intelligence*, 40; Michael Behe, "A Catholic Scientist Looks at Darwinism," in *Uncommon Dissent*, 147.
101. Lawrence S. Lerner, "What Should We Think about Americans' Beliefs Regarding Evolution?" *Skeptical Inquirer*, November/December 2005, 60.
102. Kurtz, "The Growth of Antiscience," 261.
103. Ralph Waldo Emerson, "Quotations and Originality," in *Letters and Social Aims* (Boston: J. R. Osgood & Co., 1877), 177.
104. See James Wilsdon and Rebecca Willis, *See-through Science: Why Public Engagement Needs to Move Upstream* (London: Demos, 2004).

Index